SEX ROLES

Sex Roles

BIOLOGICAL, PSYCHOLOGICAL, AND SOCIAL FOUNDATIONS

Shirley Weitz

The Graduate Faculty

New School for Social Research

New York

OXFORD UNIVERSITY PRESS

1977

First printing May 1977
Second printing August 1977

Library of Congress Catalogue Card Number: 76-55508

Printed in the United States of America

To Leon

Contents

PREFACE

How is this book different from all other books in the burgeoning sex roles-women's studies area? Three features make this book distinctive and in fact motivated the writing of the book in the first place. I have taught undergraduate and graduate courses in sex roles and covered the topic in more conventional psychology and social psychology courses for several years and have long been frustrated by the shortcomings of the books available. In writing my own book I have sought to make these three points the focus of my concern: an interdisciplinary emphasis; an awareness that no one of these disciplines provides "the" answer; and an approach acknowledging that female and male sex roles must be studied together as elements in a sex role system.

1. *Interdisciplinary emphasis.* The person is at the center of a field of biological, psychological, and social determinants of behavior. Information from all relevant fields of knowledge must be presented to make possible a full picture of sex roles in operation. As a psychologically trained social psychologist my first impulse is to see the person as the locus of causality for behavior, but my next impulse is to recognize that the person functions in a complex social world that affects behavior every step of the way. In the area of sex roles, it makes no sense to ignore the role of the family system, symbolism, history, or literature, or the media. And, even though it has unfortunately formed the basis of many a reactionary position, the input of biology cannot be swept away. Even though no book of this size can fully cover all areas, some mention of their importance should be made lest the student be misled into thinking that just because the topic of sex roles is presented within the context of a psychology or sociology course all other approaches can be safely forgotten. At the end of each chapter further readings are suggested for gaining deeper knowledge of the approach just discussed. Footnotes, of course, also supply additional sources.

2. *No easy answers.* In accord with the interdisciplinary approach taken, I believe that no one input, be it biological, psychological, or social, can be seen as "the" answer to so complex a problem as sex roles. I can offer no simplistic solution to the "why" of sex roles. I can only offer a multitude of "solutions." All sex role behaviors are multidetermined. The person lives simultaneously

within a biological, psychological, and social world at a given point in historical time. All factors exert their influence and to give priority to one over the other seems like a useless academic exercise in analysis. The real world rejects all such simplistic single-factor, cause-and-effect solutions. For example, one might single out the family as a key force in maintaining sex roles. Yet the family itself is a complex institution, having biological, psychological, economic, historical, and cultural roots. I have therefore opted to talk in terms of "maintenance systems"—inputs from the biological, psychological, and social world that simultaneously determine the nature of the sex role system. And it is a *system*, involving both female and male roles, that brings me to my final point.

3. *Female and male roles as part of a sex role system.* Many books in the women's studies area treat the position of women apart from that of men, talking in terms such as "the psychology of women." While no one can fault the treatment of so long-neglected an area like women, pretending that knowledge of the female sex role is complete in and of itself is an exercise in self-delusion. Female and male roles work in tandem as a system and must be understood in that light. Efforts at sex role change have often foundered at that very point: when female roles were changed but male roles were not, things remained the same and the traditional system was maintained. This book seeks to understand the simultaneous, interlocking nature of the roles as a way of understanding their operation and of exploring possible avenues for sex role change.

Sex Roles, then, attempts to guide the reader through the complexities of the sex role system, approaching it from a number of disciplinary perspectives, but offering no easy answer to the "why" or "how" of its operation and persistence. It is my hope that this book will be useful in fostering understanding of the system as a step toward introducing sex role change.

I would like to thank the many people who through the years have stimulated my thinking about sex roles and about life in general. The Graduate Faculty of the New School for Social Research provided the sabbatical leave necessary for writing this book, and I thank them for it. I would also like to acknowledge the able research assistance of Mary Anglin and the typing skills of Barbara Gombach. The fine people of Oxford University Press have again made the details of writing and publishing a pleasant experience. I would especially like to acknowledge the encouragement and support of

Bill Halpin, Vice President of Oxford University Press. The able copyediting of Betty Gatewood added much to the book, as did the skills of Ellie Fuchs, Virginia Watson Rouslin, and the anonymous others at Oxford. Finally, but first, I would like to thank my husband, Leon Lazarus, for his encouragement and love.

S. W.

New York, N.Y.
January 1977

SEX ROLES

INTRODUCTION

The lesson of history so far is that women cannot gain equality regardless of the methods used to obtain it.

William L. O'Neill
Everyone Was Brave[1]

Or, to put it more generally, the lesson of history is that women's roles do not change, and neither do men's. Except for a few technological innovations, women and men are performing just about the same functions within the family and society as they were in the middle ages, or in prehistory, for that matter. The continuity extends across space as well as time. Thus women in the depths of African jungles are doing roughly the same sorts of things with their time as their sisters in suburban New York, give or take an electric dishwasher or two. Both bear the primary responsibility for child care and for maintaining a household, and probably do not or *cannot* play a significant role in the economic and political structures of their societies, even though they may well contribute to the economic subsistence of their families. Both are viewed as less aggressive than men and probably as more emotional and moody. It is likely that they view themselves in these ways as well.

Perhaps even more so than for women, men's lives have changed but little. Aggression and assertion remain important core traits for masculinity and are exercised in the societal realm in political participation, warfare, and work. For the most part, the private, familial domain is female, while the public, social one is male. This pattern has been largely unbroken by history or geography. Even in societies with significant female participation outside the home, such as the United States and China, the power structure in the society remains male, while the female contribution is private and individual and is often tied to the family.

In the 1930s, Margaret Mead[2] studied three cultures in New Guinea, the Tchambuli, Mundugumor, and Arapesh, in which she claimed that sex roles were assigned very differently from those in Western culture. The Tchambuli were described as reversing sex roles, producing strong, dominant women and nurturing, "maternal" men. The Mundugumor were said to favor a hostile, supermasculine character type for both sexes, while the Arapesh were seen as emphasizing nurturance and feminine traits for its members, regardless of sex. This study has been cited time and again to demonstrate the plasticity of human sexual nature and to argue against any innate or unchangeable component in sex role assignment. However, Fortune[3] made another visit to these societies, and his findings did not agree with Mead's. He argued that even in the Arapesh society, males retained the ultimate power in the society and were solely responsible for the waging of war and other organized aggression. Fortune also disputed Mead's finding of similar sex roles for both men and women in Arapesh socieity.

Of course there have been significant changes in the definition of sex roles, particularly the female, in our own society and in others, notably the Chinese, Russian, Israeli, and Scandinavian. As we shall see, however, even in these innovative cases, the defining characteristics of sex roles have remained remarkably consistent, even though some external changes seem evident. And even where female roles have become a bit more flexibile, male roles have generally changed little, if at all. The sex role *system* has not changed though the climate may favor more *individual* latitude than before. By and large, in our own society, sex role standards and patterns have successfully withstood attempts at change. When individuals act out of sex role it is they who are regarded as deviant, not the sex role standard, testifying to its tenacity and staying power. The typical life pattern of men is generally quite different from that of women, and it is the goal of this book to examine the roots of such differences. If we can sum up the purpose of this book in one sentence, it would be to discover why we can predict so much about people's lives just by knowing one fact about them: their sex.

The essential continuity in sex roles across time and space has been used as a major indirect argument for a strong biological basis for sex roles. Thus, it is argued, all societies could not have independently "invented" the same sex roles. If no biological predisposition existed and random assignment were the rule, why not an equal split in societies, with power, aggression, and economic and political responsibility being just as often found in women as men? Also, they go on, the relative ease of socializing children into adult sex roles can be explained by innate biological and temperamental dispositions that separate the sexes. Because relatively few boys and girls rebel against the roles assigned to them, even in this day of feminine (and masculine) protest, it is argued that these roles have an innate component. Even if one were to accept this line of indirect argument, it does not lead us *only* to a biological basis for sex roles. Other *maintenance systems* can also be posited to explain such stability, particularly psychological and social systems. Within the individual, resistance to change could be accounted for by early socialization experiences, which stamp in personality traits congruent with adult roles. Society also has a vested interest in sex roles insofar as they permit the smooth functioning of the major institutional structures, such as the family, the economy, and the state.

The seeming stability and continuity of sex roles, then, argues for the existence of maintenance systems: biological, psychological, and social. It is these maintenance systems that we will consider in this book. Does study of such systems imply a belief in the status quo? Decidedly not. It is my profound belief that any program for sex role change—which I believe *is* necessary—must rest on a solid understanding of the mechanisms that have kept the current structure going for so long. "Conspiracy theories" of history, which seem to imply that men have kept women down over the centuries through some collective act of will, do not merit serious consideration. Both the individual and society are very sophisticated organisms, and any system that has remained so long in either, such as sex roles, deserves a more complex and coherent explanation. It is the thesis of this book that the three systems of biology, psychology, and society have maintained sex roles in their current form, and therefore it is at these three targets that any intelligent plan for change must be aimed. We must understand the sources of the system and the nature of its tenacity before changes can be effectively planned. Each of the following chapters considers the nature of one of these maintenance systems and indicates goals for change and already existing trends toward change. The final chapter will tie together these arguments with additional thoughts on the prospects for change.

This book embodies a personal view of sex roles. No attempt has been made to summarize the vast literature that has been accumulating in the field over the past few years. Rather, I have tried to take an integrative approach, sifting through the material with a definite perspective in mind. That perspective is social psychological, emphasizing the need to consider both individual factors and societal structure in seeking any satisfactory explanation for the origins of sex roles, the maintenance systems that keep them going, and the problems in introducing sex role change.

The social psychological perspective is a difficult one to hold to consistently, for there is always the temptation to consider only the psychological or only the societal, and to fail to see the two as an interlocking unit. Thus, it is impossible to think of individual development apart from the context of the culture and the particular historical moment, or to think of the operation of the social structure apart from the character of the individuals who make it up. Sex roles provide an excellent illustration of the application of this method, since sex roles are simultaneously part of the individual and part of the social structure.

This book will try to hold the two perspectives in mind simultaneously in an effort to find a satisfactory explanation for the existence and persistence of sex roles. The book will begin with the consideration of possible biological inputs into the sex role system. Although many would minimize the contribution of biology in our complex technological society, it does not make sense to suppose that roles would have evolved that are entirely out of sorts with biological predispositon, nor that they could endure for so long without any sort of biological base whatsoever. In this regard, two areas of inquiry seem especially worth pursuing: aggression and sexuality. Both these behaviors are to some degree sex-typed in the individual and in the operation of the social structure.

Next, we will turn to socialization, emphasizing processes that lead to the internalization of sex role standards and the development of differential motivational and skill complexes that feed into the societal structure of sex roles. Brief consideration will be given to the question of differential parental treatment of children as a function of sex, but most thoroughly explored will be the concepts of identification and modeling, which in the author's view form the core of sex role identity. Of course it is important to recognize that parents are not the only agents of socialization, and that schools, peers, the mass media, and the child's general percept of the social world all support the development of sex role identity.

At the crux of the question of sex roles is the issue of the family. The next chapter will consider the critical role played by family structure in the origin and maintenance of sex role systems. The sex-typed structure of the work world is very much a part of the whole question of the nature of the family, and it will be considered along with it. Other societal support systems will also be considered, such as symbolic, religious, and cultural processes. Finally, the history of attempts at sex role change, both within our own culture and in natural "experiments" such as the Israeli kibbutz, is very instructive. Some of the difficulties encountered, as well as the areas in which change has occurred relatively rapidly, will be examined in the concluding chapters. These areas of resistance and ease of change give us clues to the structure of the sex role system.

Throughout, several themes emerge. First, the emphasis will be on *both* female and male sex roles, since they operate as a system and one cannot be understood or changed without the other. Recently this simple fact has been recognized in the publication of several books on the nature of masculinity, the growth of movements aimed at changing the male role, and the recognition by women's groups that such a change is a necessary concomitant of change in women's role. Analyses of the failures of some sex role "experiments" in the socialist countries, such as Russia, and Scandinavia, have often recognized that it was the failure of the men to adjust to the changes and to consider the need to institutionalize them—for men as well as women—that led to the return of the old system.

A second theme will be the systemic nature of sex roles, embodying both psychological and societal components as inseparable parts. Thus, any personality differences between the sexes would lose significance if there were no sex-typed social niches set up to accommodate such differences. Likewise, the sex-typed nature of work and family roles, as well as political and historical ones, could not be maintained were it not for inputs from the psychological system, providing individuals socialized into such a setting.

Third, we will attempt to gain as much cross-cultural and historical perspective as possible on the issues involved. It is all too easy to put on blinders and assume that one's own cultural and historical assumptions are universal. Along with this breadth of perspective comes the need to consider the outcome of efforts at change, as ways to pinpoint the structure and vulnerabilities of the system, as well as to take a humanistic perspective on ways to make human life a more pleasurable experience for all concerned.

SUGGESTED READINGS

GENERAL REFERENCES:

Helen S. Astin, Nancy Suniewick, and Susan Dweck. *Women: A Bibliography on their Education and Careers*. Washington, D.C.: Human Service Press, 1971 (p).* Summaries of key studies and books in the area.

Helen S. Astin, Allison Parelman, and Anne Fisher. *Sex Roles: A Research Bibliography*. Rockville, Md.: National Institute of Mental Health, 1975 (p). Available through U.S. Government Printing Office, Washington, D.C. Extensive summaries of recent research in psychological sex differences and societal sex roles.

Simone de Beauvior. *The Second Sex*. H. M. Parshley, trans. New York: Random House, Vintage Books, 1974 (p). Classic work, integrating psychological, literary, and historical sources. De Beauvoir's autobiographical works such as *Memoirs of a Dutiful Daughter* (New York: Harper & Row, 1974) also give valuable insights.

Sue-Ellen Jacobs. *Women in Perspective: A Guide for Cross-Cultural Studies.* Urbana: University of Illinois Press, 1974 (p). Excellent comprehensive bibliography on women around the world (not annotated).

Naomi Lynn, Ann Matasar, and Marie Rosenberg. *Research Guide in Women's Studies*. Morristown, N.J.: General Learning Press, 1974 (p). Guide to sources useful for the researcher: films, periodicals, government agencies, as well as books and bibliographies.

Joseph H. Pleck and Jack Sawyer, eds. *Men and Masculinity*. Englewood Cliffs, N.J.: Prentice-Hall, 1974 (p). A collection of articles, written mostly from a personal standpoint, on the meaning of masculinity in our society.

Marie Barovic Rosenberg and Len V. Bergstrom, eds. *Women & Society: A Critical Review of the Literature with a Selected Annotated Bibliography*. Beverly Hills, Calif.: Sage Publications, 1975. Comprehensive, annotated bibliography on sociology, political science, history, and economics. Briefer treatment of psychology, anthropology, philosophy, religion, health, literature, and the arts.

Michael S. Teitelbaum, ed. *Sex Differences: Social and Biological Perspectives.* New York: Doubleday, Anchor Books, 1976 (p). Good collection of essays covering social and biological factors in sex roles.

---

*(p) means available in paperback.

PERIODICALS:

Feminist Studies, 417 Riverside Drive, New York, N.Y. 10025. Feminist analyses of important theoretical and empirical issues; multidisciplinary.

Psychology of Women Quarterly, published by Human Sciences Press, New York. A new journal sponsored by the Division of the Psychology of Women of the American Psychological Association. It features empirical work, theoretical articles, and book reviews.

Sex Roles: A Journal of Research, published by Plenum Press, New York. Emphasizes psychological research on sex roles.

Signs: Journal of Women in Culture and Society, published by the University of Chicago Press. Interdisciplinary journal with emphasis on historical, political, and economic factors in the lives of women.

Many other professional journals also feature articles or occasionally entire issues on the topic of sex roles.

LITERATURE:

Aristophanes. *Lysistrata*. Douglass Parker, trans. New York: New American Library, 1964 (p). The war between the sexes has not changed much in the intervening years.

Louise Bernikow, ed. *The World Split Open: Four Centuries of Women Poets in England and America, 1552–1950.* New York: Random House, Vintage Books, 1974 (p). The best anthology available; insights into the female (and male) condition.

Evan S. Connell, Jr. *Mrs. Bridge* and *Mr. Bridge*. Greenwich, Conn.: Fawcett, 1958 and 1969 (p). Twin portraits of a female and a male in conventional American roles; very well done.

George Eliot. *Middlemarch*. 1872; rpt. New York: New American Library, 1964 (p). Classic study of a woman constrained by societal definition of her role.

Sue Kaufman. *Diary of a Mad Housewife*. New York: Bantam Books, 1967 (p). Biting portrayal of the "successful" American housewife.

Doris Lessing. *The Golden Notebook*. New York: Bantam Books, 1973 (p). Lessing's best book on women's integration of love and work.

Women and Literature: An Annotated Bibliography of Women Writers. Available from Women and Literature, Box 441, Cambridge, Mass., 1976 (p). Biographies and evaluations the major works of twentieth-century writers, as well as some eighteenth- and nineteenth-century ones. Includes a subject index which guides the reader to books on, for example, adolescence and marriage.

1 The Biological Maintenance System

The oldest arguments against change in sex roles are biological. Birth as a male or a female is a biological fact, immutable and final. From this incontrovertible difference are deduced other differences, with implications for the individual and for society extending far beyond the primary sexual differentiation. Thus, it is biological fact that only women can bear children, but not that they alone can raise them or that only they possess personality traits congruent to motherhood. Parsimony would argue for such a biologically simple package of traits and roles, and this is usually the line of argument taken by critics of social change. However, to determine the nature and extent of biological maintenance systems in *humans* requires more than appeal to theoretical simplicity and elegance. Biological facts, insofar as they exist for humans, must be sought. This is a difficult task, since much research has been on animals and, even then, has only indirectly addressed the sex role issue, but we will attempt here to summarize what is known.

For the sake of convenience, we will consider two broad areas thought to have both strong biological roots and important consequences for sex role assignment and behavior: aggression and sexuality. In conclusion we will study the case of psychosexual abnormalities as a way of understanding the effects of sex hormones.

AGGRESSION

The dimension of aggression-passivity is a critical one for sex role differentiation. Almost all the sex-role specific behaviors can be seen as related to aggression in some way. Thus sex differences in achievement, personality, social role, etc., could be based on a substrate of differential aggressiveness between males and females. Determining the extent of such differences and their biological origin thus becomes a matter of first priority. Before noticing that men hold power in all known societies, let us turn to the possible biological origins of that power in the individual male or female. It is only on the common biological substrate of man (and woman) that society can effectively build institutions and individuals shape personal destinies.

THE BIOLOGICAL BASIS OF AGGRESSION

Konrad Lorenz[1] has put forth a controversial theory of aggression by positing an instinctual basis as an explanation for its pervasiveness and tenacity in human and animal populations. However, he does not posit a biological mechanism of action for this instinct, although, presumably, nervous and endocrine functions would be involved. Nor does he spend much time on the sex differential in aggression ("favoring" males in almost all species), which, of course, could suggest a mechanism of action for the instinct. By turning to this issue, we might perhaps discover some findings relevant to aggression in general, as well as to its specific application to sex roles.

Human sex differences in aggression are so taken for granted that we do not even stop to think that the problem of aggression in the human species is really that of aggression in *males.* War is a male game; for the most part so is violent crime. Even in suicide, women choose the nonviolent sleeping pill, men the gun. The legend of the Amazons celebrates the military prowess of women, but clearly as an exceptional occurrence. In fact, the word "Amazon" means breastless in ancient Greek, thus denying the femaleness of these aggressive women. In legend, these women were said to remove their right breast to facilitate drawing a bow, to live apart from men except for breeding purposes, and to kill their male infants.

Apart from legend, there appears to be little historical foundation for supposing that such a race of women ever existed. Some armies (of men) have been accompanied by women bringing food and ammunition; this seems to have been the case among the early Germanic tribes, the Mongol armies of Genghis Khan and the Hazaras of Afghanistan. As for having women in an actual fighting role, only a group in Dahomey (in Africa) is said to have done this. The Spanish explorer Francisco de Orellana reported a race of fighting women in South America and named the great river Amazon after them. In 1971, a Brazilian anthropologist, Altair Sales, and a German ethnologist, Jesco Van Putt Rames,[2] reported the discovery of some cave drawings and rocks they claim are artifacts of the Amazons. However, the evidence is extremely sketchy: a few triangles linked historically to Amazon relics, and drawings of flutes to lure men to mating rites. (To this day the playing of flutes by Indian women is forbidden, according to Sales, to prevent women from returning to Amazon ways.) Several rocks have also been found, purportedly the sites of ritual matings and male infanticides. The evidence is quite circumstantial and it seems safe to conclude that if Amazons really existed for any length of time in Brazil or elsewhere, we would not have so much trouble in proving their historical reality. The persistent legend of the Amazons may have more to say about male fears of female power (see the discussion on witchcraft in Chapter 4) than about the actuality of their existence.

It seems safe to conclude, then, that institutionalized aggression at least (e.g., war and violent crimes) is a male prerogative in all societies. Even modern-day Israel drafts women only for noncombat roles. What of the question of individual aggression, or aggressiveness-dominance as a personality trait? Is it also sexually dimorphic, as are the social projections of aggression? Clearly, the personality attributes most closely allied with aggression, such as dominance, independence, and activity, seem to have connotations of masculinity, while their counterparts on the nonaggressive side, submission, dependence, and passivity, arouse ideas of femininity. How far these terms go in describing individuals is another issue, and there is considerable variability, with much indirect (e.g., verbal) aggression expressed by women. Sex differences in personality do tend to point up a dichotomy along the aggression dimension. Of course, the considerable social pressure against direct expression of aggression in women must also be considered in assigning any final weight to constitutional factors.

Given this circumstantial evidence for sex differences in human aggression, let us turn to possible biological sources: brain mechanisms, hormones, and chromosomes.[3]

BRAIN MECHANISMS OF AGGRESSION

There is no single "aggression center" in the human brain. In our species, aggression is such a complex response that both the most primitive and most highly developed areas can be involved. Primitive rage reactions, akin to those in lower animals, may provide the motivational basis (or "go" mechanism) for aggression, but higher cortical areas are surely involved in the execution of such responses. Thus the Nazi racist philosophy used to justify some of the most incredible aggressive campaigns yet devised came from the same cortical areas of the brain as those capable of creating the most exquisite lyric poetry or of making out a pedestrian income tax return. For that matter, so might the plans for hydrogen bombs or a premeditated murder come from these same areas of the brain. Yet we cannot dismiss lightly the idea that so-called "lower" brain areas may be providing the motivational impetus for aggression, just as they do in animals, and that our more advanced brains allow us to add intellectual icing to the cake of aggression but not to modify its more primitive base.

For many years, the rhinencephalon or "smell brain," one of the evolutionarily most primitive parts of the brain, was virtually ignored in humans and considered to be concerned only with olfaction, a sense that plays a relatively minor role in human life. Recent work with both animals and man has indicated, however, that these structures of the *limbic system,* as we now call it, are involved in emotion and motivation, as well as in certain kinds of learning. For example, in cats, when all endocortical tissue is removed, leaving only the limbic system and other related structures, the animals still retain their characteristic emotional reactions.

Most important for our concerns is the fact that aggression seems to be an important feature of the *amygdala,* one of the structures of the limbic system. An early study, done by Heinrich Klüver and Paul Bucey,[4] revealed an unusual syndrome of behavioral reactions after removal of the temporal lobe including the limbic system and amygdala. Normally vicious rhesus monkeys became placid after this operation. Subsequent work seems to have localized the effect in the amygdala. Stimulating the amygdala through electrodes, on the other hand, seems to result in rage responses, although fear and placidity also have been reported depending on the specific area stimulated. Certain parts of the *hypothalamus,* a related structure, also produce rage reactions when stimulated in cats and other species.

The localization of aggressive behavior is a potentially important finding for sex role research, as sex differences in the structure and function of the amygdala could be explored. The amygdala seems to be quite central to aggression in humans as well as in animals. Research on habitual violent criminals (males) has led to the discovery of abnormal EEG patterns in the temporal lobe limbic structures as compared with EEGs of nonviolent criminals. This could indicate unusual neural activity, or, at least, a different sort of neural activity in violent individuals. A particular kind of epilepsy, called temporal lobe epilepsy, can lead to attacks of impulsive and violent behavior. Destruction of the amygdala in such patients leads to symptomatic relief.[5] The usual procedure is to introduce a substance into the structures (through stereotactic surgery) that will lead to disintegration of the affected structure. Neurologists using this technique report a high cure rate. Both male and female patients seem to be predisposed to such attacks and respond well to this treatment. One female patient, treated by Mark and Ervin,[6] had her amygdala stimulated through radio electronic contact and erupted into a violent burst of behavior (she did not know she was being stimulated). Later she was treated through stereotactic destruction of her amygdala, and this stopped her seizures and aggressive attacks.

Additional information about the amygdala is quite revealing. It seems to play an important role in sexual behavior in some animals, affecting the timing of puberty and ovulation. It seems to be particularly involved with the endocrine regulation of sexual behavior. There is also some evidence that the amygdala may act in some cases to inhibit prolactin secretion. Prolactin is the hormone most closely associated with maternal behavior, so that its inhibition by a center concerned with aggression is a particularly arresting fact.

Raisman and Field think that the junction of the amygdala and the preoptic area of the hypothalamus is particularly sensitive to androgen, the male sex hormone. They sat "it is tempting to speculate that it may be this connection of the preoptic region which is affected by neonatal androgen and is therefore sexually dimorphic."[7] We will soon consider the postulated organizing effect on the nervous system that such androgen may have. Raisman and Field later found sex differences in the preoptic area of the rat in the number and type of synapses reported.[8] The preoptic area is implicated in the production of gonadotrophins, substances that affect the secretion of gonadal hormones. Thus there may be an anatomical and functional connection between the amygdala and sex hormone production, a fact that could be of tremendous importance to sex role research. Along this line, John Money has reported that temporal lobe dysfunction may be related to sexuality, particularly in regard to the telling of "dirty stories," a trait that seems to be more characteristically male in our culture.

As mentioned earlier, the limbic system in man does not function in splendid isolation from the so-called "higher" neocortical areas. The connection between the limbic system and the outside world must come about through higher sensory, thought, and memory centers. Possible sex differences in cognition, as well as differential memories imprinted by male or female socialization, could interact in myriad ways with the limbic system and its associated hormonal activities (see below). Arnold[9] has speculated on the possible role of the limbic system in affective memory, in which it would work with memories stored in the cortex in devising strategies of action. Far from depreciating the role of the limbic system, as earlier theorists did, recent researchers tend to give it a directive function in relation to the cortical areas. Thus Smythies remarks:

> The cortex also feeds the latter [limbic system] with condensed indications of cortical activity, including categorized representations of the state of the external world. The limbic system constitutes the *meta-organizational system.* This appraises and evaluates the activities of the first system and balances current priorities with regard to the short-term and long-term needs of the organism, and the selection and evaluation of different integrative activities.[10]

THE HORMONE-BRAIN-BEHAVIOR SYSTEM AND AGGRESSION

Any discussion of the sexually dimorphic brain mechanisms for aggression relies heavily on hormone influences.[11] Gonadal hormones, which are already known to differ between the sexes, are thought to enter the brain and cause subsequent behavioral differences. This hormone-brain-behavior system is a relatively recent discovery and has led to the founding of a new medical speciality: neuroendocrinology, which focuses on hormone-brain interactions. For a long time, the pituitary (or hypophysis) was thought to be the "master gland," sending out gonadal stimulating hormone (GSH), which caused the target gonad to produce androgen (in males) or estrogen (in females) and other associated hormones through a feedback loop. Then it was found that the pituitary was subject to still higher brain control, localized in the *hypothalamus,* an area associated with the limbic system (see above). Neural messages from the hypothalamus were sent to the pituitary, resulting in stimulation or inhibition of the target gland (in this case the pituitary).

The most important thing to note about this system is the idea that at a very early developmental stage, the hypothalamus becomes indelibly "sex-typed" through the action of sex hormones, thereby permanently predisposing the animal to male or female physiological and behavioral responses. In most animals, this "critical period" of hormone action is thought to occur prenatally and thereafter is immutable. The "natural" predisposition is to a female system. This natural tendency can only be broken by the action of androgen, the male hormone. If androgen is present, a male develops. If not, a female. Estrogen, the female hormone, does not seem to have any crucial differentiating function. Thus, the hypothetical mode of action is as follows: for a male, androgen acts on the hypothalamus and converts it to a male, acyclical rhythm of GSH release. If the action of androgen is stopped in a genetic (XY) male by the introduction of an anti-androgenic agent, or by removal of the testes, a female physiological and behavioral system develops. A human psychosexual abnormality, testicular feminization, occurs when the androgen produced prenatally in a genetic male has no effect on bodily tissues and a phenotypical female is thereby produced, usually with female behavioral traits and female upbringing. In rats, the "critical period" for androgen action does not occur until after birth. Hence, if male gonads are removed and female hormones substituted, female behavior, particularly sexual behavior, results. Likewise, if genetic female rats are gonadectomized and given male sex hormones, male behavior is produced. The important thing to note is that the playing out of this behavior comes in adulthood, so that the initial hormonal treatment is critical. The timing of this treatment is also crucial; if it occurs even a short time after this critical period, the typing will already have occurred and will be irreversible. Interestingly enough, estrogen, which is chemically similar to androgen, can sometimes have similar effects in changing the direction of development during this period.

The implications for aggression in the workings of this hormone-brain-behavior system are clear. The permanent changes induced in behavior often involve aggression. Thus, female animals treated with male hormones during this critical period behave more aggressively in adulthood than do normal females (although usually less so than normal males). In humans, the adrenocortical syndrome results in the overproduction of prenatal androgen in a genetic female with masculinizing effects, including more tomboyish behavior (although the measures are not totally convincing; see later discussion, p. 48).

In most species, male sexual behavior involves more aggression than does the female; therefore effects on the sexual system are often intertwined with changes in aggressive behavior. Thus, if female rats treated with male hormones show more mounting behavior and less lordosis (arching of the back as a receptive sexual posture), this can also be interpreted as a change toward more aggressive, action-oriented behavior if seen in a nonsexual context.

Goy's[12] work with rhesus monkeys illustrates the application of the critical period hypothesis to aggressive behavior. In rhesus monkeys, the critical period for sexual differentiation comes before birth; testosterone was therefore given to the monkey mothers, and the resulting genetic female offspring were pseudohermaphroditic (in their genital appearance they would be judged as males). In their sexual behavior, they also resembled males. Three types of social behavior: social threat, play initiation, and rough and tumble play, were chosen as nonsexual criterion behaviors, as they are known to differ between normal males and females (males evidence more of the above three behaviors than females). Without any additional hormonal treatment, the masculinized females were found to resemble normal males in the frequency of these behaviors. Thus the single hormone treatment during the critical period of development permanently predisposed these female animals to a male behavioral pattern of response.

This irrevocable, hormonal "sex-typing" of the nervous system (which is normally, of course, in accord with the genetic sex of the animal) has the most far-reaching implications for sex differences in human behavior. If, indeed, there is this indelible stamping-in of "male" or "female" neural circuits mediating social behavior at a very early age and throughout life, then society's sex-typing is only superimposed on these natural predispositions and is not a sovereign influence in itself. Thus far, the only human evidence we have comes from the "natural" experiment of psychosexual errors and this is equivocal. (We shall consider the case of the adrenogenital syndrome later in this chapter.) Some researchers cite cases of persons raised contrary to their biological sex who conform very well and feel no strain, and others, with the same disorder, who cannot adapt and assume the behavior of their natural (rather than socialized) predisposition. Some cases of transsexualism, in which the person wishes to change his sex and live as the other sex, may be traceable to some early effect of hormones on the nervous system, although this is not established. Some recent work with male homosexuals indicates that they have a lower amount of the male sex hormone than heterosexual males, although the order of cause and effect here is not clear.

The fact that sex hormones seem to have a direct effect on aggression in all species and that they have an early and pervasive influence on the nervous system seems to be a strong factor in explaining the pervasive sex differences in human aggression. Castrated males of all species are more docile and tractable, and the effect must come from loss of most of their male hormone. Human females treated with androgen for certain clinical disorders (such as breast cancer) become more active and aggressive and more openly sexual. Aggression in male animals seems to increase at the time of maturation of the gonads, when hormonal systems are fully functional. Estrogen derivatives have been used to quiet aggressive behavior in humans.

It would be very easy to say that human socialization practices simply follow the lines of natural hormonal predispositions, as they seem to in the animal world. Indeed, in male babies, more activity (perhaps a forerunner of aggression) is apparent at even a few days of age. Temperamental differences between the sexes are recognized in almost all societies and are usually encouraged in both overt and subtle ways. Mythical systems of organizing the world, such as the Chinese yin-yang, recognize opposing active (male) and passive (female) principles. A neat little biological argument seems able to explain all sex differences in aggression and related behaviors. But, before accepting this case too uncritically, one should question the evidence, which is quite complex and is disputed even in the studies of animals. The action of sex hormones on bodily organs has *not* been fully explored in humans, nor has their effect on neural structures and complex behaviors been established. We do not know much about critical periods in human development, and we cannot say with certainty how they interact with cultural influences. For the human being, culture is part of biology, part of the stream of life he is imbedded in and influenced by. What it means to be male or female in his culture will interact with his biological heritage to produce an individual identity, making it difficult to extricate direct cause-and-effect relationships.

It is clear, then, that the hormone-brain-behavior system is one of the sources of sex role differentiation, but the extent and degree of immutability of its effect in humans is as yet simply unknown.

CHROMOSOMES AND AGGRESSION

Recently a number of cases of hyperaggressive males with an extra Y chromosome have been reported in the press.[13] Normally a male will be characterized by an X and a Y sex chromosome, while a female will be XX. The presence of the "Y" is sufficient to produce a male. In cases where Y is lacking, such as XX (the normal female), or XO (only one X), the individual develops as a female. But one Y chromosome with any X (or X's in abnormal cases) will always lead to maleness; thus, XY, XXY, and XXXY individuals are male in appearance. It is thought that the Y triggers the production of an androgen "bath" at a critical point in embryological time, leading to male development. Thus, nature's basic plan is female; only a Y chromosome will cause a deviation from this path to produce maleness. Now, the discovery that males with more than one Y chromosome appear to be characterized by hyperaggressiveness seems to lead to the conclusion that maleness (embodied in the Y chromosome) and aggression are inextricably linked, with more maleness (two Y's) producing even more of the "male" trait of aggression. However, a closer look at the evidence may lead us to doubt this conclusion.

The first studies of the XYY syndrome were done in prison populations and several dramatic cases of XYY murderers were reported. In one case, in France, the issue of genetic determinism in criminality was given as a legal defense. The American mass murderer, Richard Speck, was reported to be an XYY case, but this was later found to be an error. Studies in prisons turned up frequencies of XYY individuals that were higher than normal, especially among tall men. (Height seems to be influenced by the extra Y.) However, the key question here is that of the appropriate control group, as prisoners are differentially distributed among various strata and groups in society and it is difficult to know what the proper baseline sample should be. Thus, the prison studies are suggestive of some higher incidence of criminality among XYY males, but do not conclusively prove it.

Money notes that some studies indicate that XXY individuals—those with an extra "female" chromosome—may also be more likely to be found in prisons or other institutions than the normal population. It may be, he suggests, the chromosomal abnormality or the extra chormosome itself that produces a predisposition to deviancy (through the mediating effect of mental dullness and impulsivity) rather than the extra Y chromosome per se. He rightly points out the possible bias in emphasis on XYY and aggression:

> History, it would sometimes seem, has her own timetable for dictating the affairs of men: she could very well have timed the onset of nationwide, nay worldwide, controversy about genetically determined crime on the basis of XXY data, far ahead of the first XYY karyotyping of criminals. But XXY has none of the fake, supermale magic of an extra Y chromosome relentlessly dictating its possessor to be the victim of his superaggressiveness. The male stereotype in the popular imagination cannot tolerate the idea of the extra X chromosome of the female making the man who possesses it more aggressive.[14]

Thus the male stereotype may be biasing our observations and serving to reinforce that stereotype through scientific evidence, which is not independent of the expectations of the observer. However, in a study of 35 cases (most in prisons), some recurring patters of abnormal behavior have been found to characterize the XYY male. Such individuals seem to have backgrounds of behavioral deviancy in childhood, a lack of close interpersonal relationships, and a history of drifting through jobs. Their personalities are characterized by impulsivity leading to violence, rather than aggressiveness per se; a bisexual or homosexual orientation; excessive daydreaming; and unrealistic future expectations. Many come from broken homes, and this fact combined with all of the above pathology makes it difficult to find a neat cause and effect relationship between the chromosomal and behavioral disorders. In addition, many of the prisoners were incarcerated for crimes against property rather than persons, and their offenses were not particularly aggressive.

Thus, the question of the relationship between XYY and aggression is still unresolved. However, the evidence seems to suggest some sort of causal mechanism at work. Reports of higher than normal testosterone levels in XYY's have been reported, as well as abnormal EEGs. A childhood case of XYY "symptoms" is suggestive, since the earlier the onset of any behavioral syndrome, the weaker is the effect of environmental influences. Money reports such a case:

> He was referred when 4½ years old because he was unmanageable at home, destructive, mischievous, and defiant. He would

smash his toys, rip the curtains, set fire to the room in his mother's absence, and kick the cat, and his 8-month old brother. He was over-adventuresome and without fear; he would climb high ladders in buildings, climb up on to window sills, several storeys high, and walk into the sea without regard to depth.

He had sudden periods of overactivity at irregular intervals lasting a few hours to a few days. At those times his face would be flushed and he would pursue his particular activity with grim intent. Between episodes he would play happily and constructively. His mother described two personalities—one considerate and happy, the other disgruntled and unstable.

He started school at 5 years, and his behavior became intolerable at times. He had a particular interest in sharp-pointed objects, and had been suspended from school several times because of dangerous and aggressive use of sharp instruments. He rammed a screwdriver into a little girl's abdomen . . .

He began wandering at 2 years. Since the age of 8 years he has become progressively less amenable to normal discipline. He began playing truant and was apprehended by police on five occasions, having travelled considerable distances by train and ferry, and once was found on the street at 5 a.m. with his small brother. The juvenile liaison officer of the police felt that his supervision and advice meant little to the boy.

His brother is normal, and his parents are a concerned, loving and intelligent couple.[15]

Of course, no one case report can be used as definitive evidence, since many instances of aggressive boys with perfectly normal XY profiles could be produced as well. To date most reports agree that "there is a definite association between the XYY genotype, and presence in mental-penal settings, but both the nature and the extent of this association are yet to be determined,"[16] suggesting an interaction between genetic and environmental factors. Since nondeviant XYY men are found, it is possible that only those subject to adverse social psychological environments during socialization may display deviant behaviors, but that the tendency to such behavior is stronger when the chromosomal abnormality is present. Recently, an attempt was made in Boston to identify XYY males at birth so that favorable environmental factors could be provided to forestall possible later development of deviant behavior. The project was stopped, however, when the community protested that such labelling would stigmatize the children involved on the basis of an unproved association.

FINAL THOUGHTS ON AGGRESSION

Evidence for a hormonal basis for aggressive behavior in humans is present but not yet compelling. Not enough is yet known of the intricacies of the workings of the brain to specify positively the ontogeny of complex behavioral responses, such as aggression, especially when an early critical-period sex-typing process may be involved. The results of animal studies to date tend to support such a possibility but do not rule out alternate interpretations, especially in the human case.

The separation of biological and cultural factors in any area is a trying endeavor; nowhere is this more evident than in the case of aggression. When societal structures and socialization practices favor the expression of aggression, it is displayed, more often by men than women. Such differences may suggest that the environment can act to bring out and enlarge upon inherent propensities, but that once established, the societal setting itself may function autonomously from the original biological roots. Besides hormonal factors, of course, well-known differences between the sexes in size and muscular strength doubtless contribute to male aggression. The adaptive significance of such differences, acting in conjunction with the nature of the family system (featuring the mother-child dyad), is another possible but partial explanation for the linkage of aggression with maleness. Female aggression is, of course, not unknown, and in animal species maternal aggression in defense of the young is often quite pronounced. However, in the realm of physical aggression and societally structured aggression (i.e., warfare), males clearly lead the field.

SEXUALITY

In this section we shall consider the nature of sexual behavior, the menstrual cycle, and the (hypothetical) maternal instinct as areas relevant to the reproductive roles of men and women, considered in their biological context. Of course, it is difficult if not impossible to disentangle the biological from the social meaning of sexuality. By and large, though, we postpone an extended discussion of the social context of the family until Chapter 3. In the case of menstruation, however, the biological and social context are so intertwined that we shall discuss both in this chapter. We shall also indulge in what some readers may think is an overly detailed consideration of the menstrual cycle. The reason for this is not so much that we believe it to play a large role in women's lives, but rather that discussion of the menstrual cycle has been so prominent in arguments over sex role change (the "raging hormonal imbalance" school of thought that was widely publicized in the news a few years ago).

The areas chosen do not constitute an exhaustive survey of the field in any case, but include those areas most relevant to sex roles. Let us begin our consideration of sexuality at the logical starting point, sexual behavior, and then proceed to the menstrual cycle and maternal behavior.

SEXUAL BEHAVIOR

A rather curious sight can be observed almost any day on a number of busy streets in New York City. A man stands on the street, and after a cursory glance at each passing pedestrian, either does or does not present the unwitting walker with a small card. A few minutes' observation ascertains that it is only male pedestrians unaccompanied by women who are given such cards. A few steps away, a woman (or sometimes a man) is going through the same performance, surveying the passing masses, but this time handing the cards only to women. The contrast in the sexual content of these cards reveals a deep split in the meaning of sexuality for males and females. The cards given to men advertise massage parlors, symbolic of sex as recreation and pleasure and the relief of insistent need apart from any entanglement with the rest of their lives. The cards offered to women, on the other hand, proffer free pregnancy testing and later abortion services, signifying the fact that for women sexuality is a much more encompassing phenomenon, difficult to separate from the reproductive cycle and entire life plan. Although there is increasing emphasis on the play element of sex for women as well as men, the socialization of women goes much further in integrating sexuality with reproduction, emotional ties, and the family, than that of men does.

The differing reproductive roles of men and women, then, cannot be dismissed as unimportant factors in the shaping of sex roles, and specifically in the shaping of male and female sexuality. Female sexual identity extends far beyond the sexual act, whereas male sexuality is almost totally encompassed within it. When we consider sexuality as a biological backdrop for socialization throughout the life cycle, we must concentrate on the different meanings of sexuality for men and women.

Historically, an important inhibiting factor in female sexuality has been the real possibility of pregnancy as a consequence of sexual activity. The vignette we began this chapter with, of males concerned with massage parlors and females with abortions, suggests this all-important distinction in the stakes men and women have in sexual expression. Fear of pregnancy has been used for a very long time as a realistic deterrent to female sexuality. This fear is now very largely a thing of the past with the advent of reliable birth control and abortion, but it still lurks in the internalized attitudes of the person and the values of society. Given the nature of the reproductive system and of currently available birth control procedures, it is still the woman who must be largely concerned with such matters and responsible for the consequences if precautions fail (i.e., abortion or pregnancy).

Extended consideration of sexuality must, of course, take into account the nature of the family and the different reproductive roles of the sexes. We shall postpone a full discussion until Chapter 3, but shall pause here to make some general observations on the nature of male and female sexuality per se insofar as they affect the nature of sex roles.

How true is the "fact" of the duality of male and female sexuality? This issue can be considered from several perspectives. First of all, is it true, as has often been maintained and as is commonly believed in the general culture, that the male sexual drive is stronger and more appetitive in nature than the female? That it is tied to a dominant, active role? The concept of appetitive behavior implies an active seeking after a goal; that is, even if the motivational object (in this case, the female) is not present, the male will go out and seek it. Most cultures certainly act as if this were the case. Middle Eastern cultures, and to a lesser degree, European ones, sequester women on the assumption that they must be protected against male sexual appetites. Seldom is the rationale given that it is females who will pursue males if they are not hidden away although some Islamic societies do have a vision of insatiable female appetites. One suspects, however, that this is a cultural rationalization rather than a report of direct reality. The existence of prostitutes and of pornography aimed at men provides supporting evidence for this conception. The physical possibility of rape is of course involved as well.

But is the male sex drive "stronger"? Or just not as subject to social regulation as the female's because of the male need to protect legitimacy and access to women? A negative response couched in a rather controversial theory of female sexuality is given by Mary Jane Sherfey,[17] a psychiatrist. She offers the counter-theory that the female sexual drive is actually stronger because of the possibility of multiple orgasms, but that society has acted to repress female sexuality because of the need to socialize women into the structure of civilized agrarian society. She bases her views on the research of Masters and Johnson, in which women were shown to be capable of multiple orgasms; on hormonal studies of early female development; and on primate studies in which the aggressive sexual drive of the female during heat is documented. According to Sherfey:

> Women's uncurtailed continuous hypersexuality would drastically interfere with maternal responsibilities and with the rise of settled agricultural economies, man's territorialism became expressed in property rights and kinship laws. Large families of known parentage were mandatory and could not evolve until the inordinate sexual demands of women were curbed.[18]

Repressive measures such as clitoridectomy (removal of the clitoris), seclusion, and severe punishment for female adultery were needed to effect the change in female sexuality. Sherfey's theory is largely inferential and does not rest on much hard evidence, however. The control of female sexuality as a measure to ensure legitimacy of births and patriarchal rights will be considered more fully in our discussion on the nature of the family.

Until rather recently, the Western cultural view of female

sexuality was not clearly articulated, although passing mention was sometimes made of the existence of female sexual desire. According to a recent study by historian Carl Degler,[19] even during the Victorian era, many physicians writing on the topic recognized the female sexual response. Other cultures, particularly non-European ones, seem to recognize the existence of female sexuality. For example, anthropological reports from Pacific tribal cultures often mention that female sexuality is a culturally recognized fact that is allowed some degree of social expression, as in adolescent liaisons. On the other hand, Western accounts of male sexuality (e.g., those in the Bible) take for granted the pressing and imperative quality of male sexual desire.

Biologically oriented discussions of the differences between male and female sexuality often emphasize the point that the male orgasm and consequent pleasure in the sexual act is a requisite for reproductive success, whereas this is not true for females. That is, for conception to occur, male ejaculation is necessary, accompanied by orgasm and pleasure as motivating states, whereas female orgasm or even (in the case of rape) passive acceptance is not required. Therefore, the argument goes, it is adaptive for the species to have pleasure and sexuality linked for the male, while this connection is evolutionarily unnecessary for the female. Making the male sexual drive appetitive, so that sexual outlets are sought, serves the purposes of widespread genetic dispersal as well as ensuring the insemination of females (since one male can inseminate many females). Evidence that males are more sexually stimulated by visual materials through distal sensory channels also supports the argument that males are set up for appetitive sexual behaviors. The emphasis on female adornment (rather than male) in most cultures would also support this view, as would the preponderance of male-oriented visual pornography. Although pornographic materials aimed at women have appeared recently in our society (e.g., *Playgirl* magazine), almost all pornographic material is still aimed at men. Societal pressures against female erotic stimuli may have played a role in this imbalance, so some future change may be expected.

It is important in any discussion of biological mechanisms to distinguish between the issue of origins and that of contemporary maintenance systems. For example, there may be some biologically induced predisposition in males to respond to visually communicated erotic stimuli, a fact taken advantage of by advertisers, fashion designers, and producers of pornography, among others. But once these institutions are themselves established in the culture they can act to enhance and stimulate the original motive in their own right.

For example, if a male grows up in a culture like our own, his "natural" propensities (whatever they might be) to respond erotically to visual stimuli are confounded by his cultural conditioning to widespread use of the unclothed female form to sell almost anything from cars to men's clothes.

Discussions about the degree of pleasure experienced in intercourse by men and women are bound to be inconclusive, since no one has yet experienced both states, though sex surveys such as Kinsey's seem to suggest that males at least *report* more pleasure. Males report more orgasms at earlier ages than women do. They report more experience with masturbation and premarital intercourse than women, although social expectations are likely to suppress female responses. Psychologists, such as Judith Bardwick,[20] think that the actual physical difference in the accessibility of the sexual organs is at least partially responsible for the greater male sex drive. The external presence of the penis, its sensitivity and erectile response, would seem to make it a natural object of pleasurable focus for the young boy. On the contrary, the clitoris is less accessible and less likely to be eroticized by the girl. The use of the penis in urination may also act to orient the boy to it. The issue of the male body image as a stimulus to psychological sex differences has been emphasized by many writers. Endless discussions of the assertive, erect male presence, as opposed to the passive, penetrated female one have dominated this area, with links then drawn to parallel psychological traits. Freud's theory (discussed in the next chapter) is perhaps the most celebrated case of this sort of thinking. Although parallels can be drawn, the plausibility of such paradigms does not establish their psychological reality for the person.

David McClelland[21] has speculated on the psychological effect of the phallic tumescence-detumescence cycle, maintaining that the association between rise-pleasure-assertion may become central to the patterning of the male personality. On the other hand, he interprets the female sexual experience as one of penetration, associated with "the often painful experiences of menstrual flow and childbirth,"[22] in preparation for childbearing and dependence on a male. The latter interpretation is certainly heavily male-centered with its negative views of female sexual functions. The interpretation of male body image as the prototype for male personality depends on the validity of the link between male sexuality and dominance. The mere fact of erection, though symbolically linked to assertion, is not enough to establish the contingency. Some primate research does indicate the use of penile erection as a threat behavior in territorial marking, however.[23]

Recent primate research has probed the association between social dominance and reproductive success.[24] The old, accepted notion was that high rank in the social dominance hierarchy is associated with reproductive success. Many primate species do seem to show a dominance hierarchy of males, maintained by ritualized aggressive encounters. Even this conception has come under fire recently, with the thought being that the dominance hierarchy was in the eye of the beholder, the ethologist and behavioral scientist used to thinking in human terms of dominance and submission as key elements in any social system. Be that as it may, recent studies have indicated that over a male's lifetime, dominance does not seem directly related to reproductive success. Nondominant males may actually copulate more frequently with females to maximize success although they may not have access during the more fertile days of estrus. Females may select males rather than the other way around, and in this case, dominance rank may be only a weak determinant of choice. In any case, primate dominance, if in fact it does exist, may not be a constant characteristic of the male. Each male through his lifetime may in fact pass through many ranks, instead of being fixed at just one, as older conceptions have maintained.

In the human case, at least in Western culture, male strength and assertiveness are qualities that are stereotyped as being attractive to women. These qualities are also highly valued by the culture, so it is hard to disentangle the effects of biologically induced preference from societally induced values. Probably both play a part. Recently, however, in American culture, there has been a turning away from high valuation of the aggressive, military model, for males, although it is probably still true that assertiveness, even in a pacifistic context, is viewed as more desirable for men than passivity.

What can we conclude, then, about the biological component of sexual response as a shaper of sex roles? Hard facts, isolated from cultural influence, are almost impossible to establish, so a few speculations are in order. There probably *is* a difference in the psychological consequences of male and female sexuality. Male sexuality is probably more appetitive and autonomous from the interpersonal context than female sexuality. Male sex hormones do seem to enhance female sexual response when administered in therapeutic contexts (e.g., in the treatment of breast cancer). In normal women, sexual desire seems to be implemented at least partially through adrenal secretion of androgen, the male sex hormone. The external nature of the male sex organ, its erectile quality, and the plain requirements of the reproductive and sexual situation probably act to promote male appetitive sexuality. Female sexuality, whatever its original nature, seems to be much more fully integrated into the interpersonal context of emotional meaning and family roles. Although a potential source of pleasure for women, this pleasure does not seem, at least for most women, to be an appetitive motivating state operating outside of the constraints of the family system.

What of the consequences for sex roles? The implications appear greatest for the family system, which we will treat at length in Chapter 3. The sexual vulnerability of women has been an argument for the seclusion of women in many cultures and for their exclusion from public life. The differing consequences of sexual activity for men and women, short-term pleasure for one, possible pregnancy and long-term maternal responsibility for the other, have doubtless largely determined the nature of societal restrictions and structures. The invention of reliable birth control devices is still a relatively recent innovation, so that the impact on female sexual attitudes remains to be determined. (Some early evidence indicates the retention of conventional female attitudes in single women, in that premarital intercourse is entered into more as an acquiescence to male sexuality and a way to maintain an interpersonal relationship than as a way to relieve sexual tension.)[25] It is possible, although not very likely, that birth control may put female sexuality into more of a male mold with far-reaching consequences for the family system and the care of children.

Finally, the issue of the confounding of male sexuality with dominance and female sexuality with passivity and submission is not an easy one to resolve. First of all, as we have seen in our earlier discussions of aggression, male hormones acting outside of the reproductive system may be enough to account for a large part of the sex difference in aggression. These sex hormones, of course, have their origin in the sexual system, and an adaptive role for them in the reproductive system is not unlikely. The requirements of the sexual act do seem to dictate somewhat greater activity for males than females, but it is uncertain how much this distinction colors the nature of other sex role relevant behaviors. In individual sexual relationships there may be a carryover to nonsexual areas, which may then shape other behaviors as well.

All sexual behavior, however, is expressed in some sort of social context, so regardless of the biologically induced constraints, we cannot fully comprehend the system until these factors are taken into account. The reader is therefore advised to withhold final judgment until the chapter on the family is considered. A look at the research on the biological basis of maternal behavior, to be discussed later in this chapter, will also be helpful.

THE MENSTRUAL CYCLE

The biological and social meanings of the menstrual cycle are inextricably intertwined and join in creating the psychological effects of the experience. In every culture, menstruation is viewed in a certain way, and this view is transmitted to its members and inevitably shapes the way a woman sees herself and experiences her menstrual periods. We will save a detailed consideration of these ideas until the chapter on symbolism, since menstrual beliefs are closely related to the mythology of birth and creation, which will be dealt with at length there (see Chapter 4). However, we can note here the general tendency of many cultures to consider menstruation as a ritually unclean state, posing a potent danger to males. For example, the Mae Enga tribe of New Guinea is reported to believe that

> contact with it [menstrual blood] or with a menstruating woman will, in the absence of appropriate counter-magic, sicken a man and cause persistent vomiting, "kill" his blood so that it turns black, corrupt his vital juices so that his skin darkens and hangs in folds as his flesh wastes, permanently dull his wits, and eventually lead to a slow decline and death.[26]

The fear of menstruation and the taboos surrounding it seem to fall into a pattern with other common cultural beliefs about the evil potential of women, which we will consider in Chapter 4.

Lest the reader think that such beliefs are confined to the highlands of New Guinea, let us pause a moment to consider our own culture's view of the experience. While ritual seclusion is not practiced, and taboos (such as not touching plants) are not institutionalized, the general cultural tone is still negatively tinged. Menstruation is not generally spoken about in public (or in private, for that matter) except with the use of elaborate circumlocutions. Illustrative of this fact was a recent episode of the popular television series, *All In The Family*, where the topic was briefly touched on, arousing much anger and annoyance in the conservative Archie Bunker, and in not a few viewers as well. Perhaps more noteworthy about this incident is the fact that that mention of menstruation is probably the only one ever noted in over thirty years of television programming. The avoidance of the topic is an indirect message to the young girl of the private nature of the experience, and perhaps most importantly, the need to shield it from public (especially male) sight. Most bodily functions that are hidden in this society, e.g., urination and defecation, take on a somewhat negative cultural meaning, and for women, menstruation is assimilated to these meanings. The effect of this secrecy may be especially traumatic for the young girl experiencing it for the first time without any advance preparation. Thoughts of hideous injury, impending death, or sexual misconduct are not uncommon, and may color the woman's later view of her entire sexual functioning, not just menstruation. Because the onset of menstruation signals the beginning of sexual maturity, many parents may be reluctant to explain the meaning of the experience fully, and leave the young girl with many lingering doubts and uncertainties. Nowadays, sanitary napkin companies often provide helpful books explaining the menstrual cycle (and advertising their products), which may provide some assistance to mothers afflicted by their own menstrual and sexual anxieties. It is probably safe to assume, however, that boys are usually not given much information on menstruation and in that case, old boys' tales are likely to proliferate and contaminate attitudes about women and their sexuality.

A clue to the pervasiveness and depth of attitudes about menstruation can be found in the use of euphemistic expressions.[27] Most women used expressions with negative references, such as "the curse," being "unwell," etc., as well as elaborate euphemisms, such as "I've got my friend," "Aunt Sylvia is visiting me." The most common male expression was "on the rag," referring to the use of sanitary napkins, but extending into a general negative view of menstruation. This phrase was generalized as a depreciation of women, particularly in regard to their emotional state. Thus, an informant said that "it describes a person who is moody or in a bad mood. If used in reference to a girl, this phrase attributes this moodiness to her period. Menstrual cramps and discomforts are known to create moodiness. If used in reference to a guy, it is highly insulting since you are saying that his moodiness is like that of a girl with menstrual cramps."

What then are the "facts" of menstruation? Since no human biological event, be it birth, death, or menstruation, occurs outside a cultural context of associated beliefs, we see that the "facts" of the menstrual cycle are social facts as well as biological ones. The attitudinal framework in which menstruation is embedded is as important to an understanding of sex roles as any study of hormonal fluctuations. However, the latter is important, too, especially since the argument about biological limitations on sex roles is so often couched in terms of hormonal determinants. Expectations about emotional states may be as important as hormonal levels in determining mood and behavior, especially when these expectations are part of the general cultural heritage as well as the socialization of women and men. We turn now to a brief consideration of the biology of the human menstrual cycle and to an understanding of what is known about the association of hormonal states and emotional states, in the context of cultural beliefs about the nature of such an association.

The average menstrual cycle[28] is twenty-eight days long (although "normal" cycles can range from twenty-one to forty days) and is divided into three parts: menstrual, proliferative, and secretory. The "purpose" of the menstrual cycle is, of course, the preparation of the uterus for the implantation of the fetus; the material expelled during menstruation is largely made up of this unused uterine lining. The menstrual flow takes up the first five days of the cycle, and is associated with relatively low levels of both progesterone and estrogen, the two key female sex hormones. There is a gradual increase in both hormones, during the proliferative phase, as well as an increase in secretion of FSH (Follicle Stimulating Hormone), which stimulates ovarian follicles to develop, with the ultimate objective of readying one ovum for ovulation. Estrogen levels are also regulated by FSH and, in turn, estrogen stimulates the buildup of the uterine wall, with the complex blood vessel system and glandular structure needed to nourish the (potentially) implanted fetus. The high level of estrogen in the follicles causes FSH secretion to decrease (through a negative feedback loop) and LH (Luteinizing Hormone) to peak. This sudden peak in LH is thought to lead to ovulation, the passage of the ovum from the follicle through the ovarian wall. Ovulation ends the proliferative phase of the menstrual cycle. This event usually occurs on the fourteenth day of the cycle, and is sometimes discernible to the woman as a slight abdominal pain ("Mittelschmerz"). It is only at the ovulation period that impregnation is possible. The effect of oral contraceptives is entirely due to their ability to stop ovulation through the introduction of competing hormonal inputs.

At ovulation, estrogen levels off and slightly decreases, but progesterone builds up to a peak. The post-ovulationary part of the cycle, the secretory phase, is largely under the control of progesterone. The primary role of this hormone is to stimulate glandular secretion in the uterine lining. Near the end of the secretory phase, both estrogen and progesterone are at relatively high levels, but then rapidly drop if no impregnation and subsequent implantation have taken place after ovulation. The uterine lining cannot then be maintained and is expelled, beginning menstruation.

Most of the attention of sex role theorists has been to these last few days of the secretory phase, the premenstrual period, when there is a rapid dropoff in hormonal level. Symptoms reported during this phase, collectively characterized as the "premenstrual syndrome," have included physical complaints such as headache, cramps, breast sensitivity, and weight gain, and psychological ones such as tension, depression, emotional lability, and irritability. It is to this latter constellation of mental symptoms that we shall turn our full attention, since they are of the most consequence to sex role development. Medical thought on the cluster of physical symptoms seems to center on explanations of water and sodium retention, engendered by the action of estrogen, although there is not universal agreement on this explanation. Considerably more controversy exists around the notion of a psychological cluster of premenstrual traits, as we shall soon see. It should be noted here that there is considerable individual variability in the report of physical as well as psychological symptoms among women, so that any biological explanation will undoubtedly have to explain such variability in both biochemical and cultural terms.

A well-known popular book on the subject, *The Menstrual Cycle,*[29] by an English physician, Katharina Dalton, takes the existence of the psychological premenstrual syndrome for granted and attributes it to water and sodium retention and sodium depletion. She cites a number of correlational studies relating accidents, crime, and school failures to the premenstrual period. Such data have of course been used to justify the exclusion of women from jobs as airline pilots and business executives, and from other positions that require skill and judgment. Therefore it is necessary to pay close attention to the sorts of studies cited to determine whether drawing such powerful conclusions is valid. Mary Brown Parlee[30] has written a very influential review of the premenstrual syndrome literature, criticizing the use of the concept and the studies used to support its existence. She notes that four types of evidence have been used in this area: correlational data, retrospective questionnaires, daily self-reports, and thematic analysis of unstructured verbal material. We shall consider each in turn, giving illustrative cases and our own and Parlee's criticisms of each.

Correlational Data

This type of data is perhaps the most damning because of its dramatic quality and ready comprehensibility to the average layperson. In this sort of study, some well-defined behavioral event (usually of negative social consequence) is related to the part of the menstrual cycle the woman is in to see if there is more frequent occurrence at one point (usually the premenstrual days) than another. Behaviors such as accidents, crimes, suicides, and mental hospital admissions have been found to have positive correlations with the premenstrual period. Parlee points out, however, several flaws in these studies. First of all, the length of the various cycle periods often is not specified, nor is the source of information as to the cycle point given. Both of these omissions could be potential sources of error and bias in the studies, especially if investigators were looking for a positive correlation and could have coded their data with this in mind. Often researchers do not report the magnitude of differences between groups as long as the sought-after premenstrual effect is found. Thus, a study by Dalton showed that 27 percent of premenstrual period schoolgirls displayed a decrement in the quality of their schoolwork, but also that 56 percent manifested no change and 17 percent improved. Later citations of her work often neglect the last two groups, emphasizing the drop in 27 percent of the sample but ignoring the lack of premenstrual effect in the majority of the group. Parlee also makes the radical suggestion that the negative behavioral event could affect the onset of menstruation rather than the other way around, thus creating a correlation that had been erroneously interpreted as meaning the opposite, that the premenstrual period led to the event. There is a clinical literature to support this assertion, in that emotional factors have been known to alter the timing of the menstrual cycle. Finally, Parlee notes that a few of the often-cited studies turn out to be anecdotal and highly unscientific when examined in closer detail. For example, the serious charge that menstruating airline pilots are accident-prone turns out to be traceable to a two-page report in a 1934 aviation journal consisting of "reports of three airplane crashes over a period of eight months in which the woman pilot *was reported to be menstruating* at the time of the crash". Without offering further data, Whitehead asserts that ". . .within the last six months there have been a number of serious and fatal accidents among women pilots and at the time of these accidents it has been found that they were in their menstrual period. . . . Some localities in the United States have been practically depleted of women pilots by accidents."[31] (Emphasis added.) One could propose that male aviation observers biased against women pilots could have provided this explanation of the accidents, independent of any corroborating evidence.

Retrospective Questionnaires

In this type of study, women are asked to check off various physical and mental symptoms occurring at different points in the menstrual cycle, and these are then correlated to the phases of the menstrual cycle (usually menstrual, premenstrual, and intermenstrual). One problem with this sort of measure is that most women know of the purported existence of menstrual and premenstrual distress as an expectation about women in the culture, and their answers might be biased to be in accord with these ideas. However, as Parlee notes, a close look at the studies themselves shows that there is no clearly defined set of distinctive symptoms at any point in the cycle, and that scores across cycle periods tend to be correlated. A British study also showed that neuroticism scores correlated with reports of menstrual and premenstrual distress, suggesting that "a woman who complains of premenstrual irritability is more likely to be irritable at other times as well, and it therefore seems as though premenstrual symptoms are an exacerbation of personality traits, which in turn are related to neuroticism."[32]

Daily Self-Reports

The principle behind this method is basically the same as for the retrospective questionnaire, that is, a self-report on a checklist of symptoms is called for. The difference lies in the fact that the reports are made on a daily basis instead of depending on recall of past cycles. Again, though, respondents are often aware of the purpose of the study and socially induced expectations about menstrual symptoms may bias the data. Research using this technique often turns up a set of widely varying symptoms associated with different points on the cycle. Parlee points out that there often is a failure to distinguish between dysmennorhea (painful menses) and other physical and psychological symptoms, so that almost any collection of symptoms has been classified as the premenstrual syndrome.

Thematic Analysis of Unstructured Verbal Material

This technique avoids the bias of the two previously mentioned self-report techniques by not supplying symptom lists to the subjects, but rather by relying on a record of their free-flowing thoughts "about any life experience" they wished to relate. Such responses are tape-recorded at various points in the menstrual cycle and then subjected to later content analysis to see if there are any systematic changes in affect that occur at critical points in the menstrual cycle. Gottschalk has developed this technique, using a content analysis scheme of categories such as hostility and anxiety. Two studies using this technique bear specific mention. Ivey and Bardwick[33] carried out a study in 1968 on twenty-six college students, from nineteen to twenty-two years of age. Subjects were asked to volunteer for a "study of the menstrual cycle," and were called in at ovulation and before their menstrual periods for ten-minute intervals during two different cycles. Basal body temperature and the timing of menses were used as the criteria for the two points. The women were asked to speak for five minutes about "any memorable life experience" and then were questioned about their moods for an additional five minutes. Transcripts of the tape-recorded sessions were then content-analyzed for various sorts of anxiety, according to Gottschalk's scheme. Results showed anxiety at premenstrum significantly higher than at ovulation. An example from one of the women will illustrate the effect. Here is a sample of speech at premenstrum, scored for death anxiety:

> I'll tell you about the death of my poor dog M. /. . . oh, another memorable event, my grandparents died in a plane crash. / This was my first contact with death/and it was very traumatic for me/ . . .Then my grandfather died/and I was very close to him. /

The next sample comes from the same woman at ovulation:

> Well, we just went to Jamaica/and it was fantastic/the island is so lush and green and the a . . . water is so blue/the place is so fertile/and the natives are just so friendly. /[34]

These results should be viewed with extreme caution, however, since all the women involved were told that they were participating in a study of the menstrual cycle and were specifically called in for responses at two key points in it: ovulation and premenstrum. It is not unreasonable to suppose that such women might consciously or unconsciously have been supplying material in accord with their own expectations about how they *should* feel at the two points, especially when reporting to psychologists who were obviously interested in this comparison. Even if some biological effect were actually present, it is likely that these expectations enhanced the naturally occurring differences. More recent studies try to avoid this problem by not informing the women about the link of the study to the menstrual cycle.

A second study using the thematic analysis technique is that done by Karen Paige,[35] published in 1971. Paige's study is of special interest because of her attempt to link premenstrual symptomatology to a specific biochemical mechanism, that of monoamine oxidase (MAO) activity. She studied three groups of women: those using combination oral contraceptives, those using sequential oral contraceptives, and those not using oral contraceptives. The women were not specifically told the purpose of the study. The two groups of women taking the birth-control pill provide important hormonal contrasts, since the combination pill provides a fairly constant and high level of estrogen and progesterone throughout the cycle (as compared with the normal cyclical fluctuation discussed earlier). The sequential pill (which has since been removed from the market because of health risks) mimics the natural actions of the body by maintaining the cyclical fluctuations while providing the same contraceptive effect as the combination pill. Paige found that anxiety increased premenstrually for the control group of non-pill-users, and also slightly for the sequential-pill users, but not for the combination-pill users, thus supporting a hormonal hypothesis. Paige hypothesized that the normal progesterone-engendered rise in premenstrual monoamine oxidase (MAO) levels was prevented by the relatively high levels of progestin in the combination pills, and that these low MAO levels maintained the positive affect level throughout the cycle. (MAO level has been shown to relate to negative affect in a number of psychiatric studies.) MAO levels were not directly measured, however, so the linkage remains a plausible conjecture, not a proven fact. A confounding variable was introduced in the study, however, by the fact that severity of menstrual bleeding was a better predictor of anxiety than hypothesized hormonal level (as inferred from pill group) and that sequential-pill users are more likely to have a heavy flow than combination-pill users. Paige suggests that menstrual distress may at least be partly caused by an anxiety response to the bleeding itself. Another interesting finding of the study was the slightly higher hostility level for the combination-pill users throughout the cycle (with no cyclic shift), which was attributed to the fact that high estrogen levels in the sequential pills inhibit MAO activity. Clearly, much work remains to be done in establishing the nature and extent of the hormonal mechanisms behind these effects.

Parlee concludes her critique by noting the bias that has resulted in more frequent publication of studies that tend to confirm the widely accepted notion of the premenstrual syndrome than those that do not find the cyclic effect. She calls for better definition of the symptoms associated with the premenstrual syndrome (over 150 have been reported in the literature!), use of several methodologies, search for biochemical mechanisms, and selection of better control groups (including men). The baseline for comparisons is also called into question, since Parlee suggests that the data found are as consistent with an explanation of a midcycle syndrome of positive traits and events as a premenstrual syndrome of negative symptoms.

A very recent study has tried to remedy some of the problems associated with menstrual studies in the past by comparing moods of women taking the combination pill with those not on the pill, using men as a baseline group. Women were not questioned about their menstrual cycle until after the study was over to prevent expectation effects from coloring the data. In addition, all subjects were questioned about positive and negative external events that had occurred to them so these could be related to mood ratings. The results were that women showed more positive moods than men in several ways. The authors slyly comment: "To our knowledge, there is no research program afoot to determine whether such dull, flat lives might be harmful to the men as individuals or threaten their capabilities to assume positions of responsibility."[36] Physical symptoms were related to the menstrual cycle for both female samples, and psychological symptoms showed that pill-taking women peaked on negative affect during the premenstrum, while women not on the pill demonstrated more negative affect during the menstrual period itself. However, when the researchers compared the correlations of mood with cycle point to the correlations with environmental stress or pleasure, they found that the environmental events and activities accounted for more of the variance in mood than the cycle point. They caution, however, that individual differences were a more potent factor in reports of mood than any of the measured factors of cycle, sample, or events. They conclude that it is likely that both physiological variables and environmental events contributed to the mood effects observed. They also note that the interpretation of environmental events might be affected by cycle point, as mediated by societally induced expectations about menstruation.

Randi K. and Gary F. Koeske[37] have put the whole premenstrual syndrome question into an attributional framework. In recent years, attribution theory has provided a viable paradigm for understanding how people view themselves and others. According to the attributional view, we tend to impute causality to almost any event or feeling state, and it is this attribution of meaning that is most important in understanding the event's impact on the individual. Given that there is a fairly well-codified set of beliefs about menstruation and its impact on personality, which is currently held by many women and which is present in the culture, it is not surprising to note that this belief system may act to shape the woman's perception of herself and her moods. In particular, there is a strong belief linkage between negative moods and the premenstrual period. The Koeskes designed a study in which undergraduate men and women were given information about a hypothetical female student, including her psychological state and outlook on life, positive or negative external events, as well as medical information that included the important variable of menstrual cycle phase (e.g., "menstrual period expected in two days" vs. "in three weeks"). Subjects tended to attribute the onset of negative moods to the premenstrum, discounting the impact of external events. The Koeskes speculate that such an attribution pattern may be a causal factor in the phenomenon of premenstrual tension. That is,

> the simultaneous operation of a cognition linking negative moods and the premenstrum could result in the discounting of situational factors and an emphasis on biology (a type of internal attribution) in explaining negative but not positive moods. This attributional pattern, then, could reflect not the actual incidence of positive and negative moods at premenstruation, but the extent to which women's negative but not positive moods are seen as explained by biology. Exaggeration of the likelihood of recall of negative moods with a reversal for positive moods might also result. Acceptance of this attribution pattern might then adversely affect women's self-esteem.[38]

Later research by Randi K. Koeske[39] showed that there was a special emphasis on biological attribution for hostility, an out-of-role behavior for females. Koeske's work certainly does not rule out a biological explanation for the premenstrual syndrome, as the results could simply reflect the widespread knowledge of a valid causal association. However, it does suggest that the attributional factor may work in tandem with whatever biological determinism does (or does not) exist, so that the causal conclusion drawn can act to strengthen the association in the woman's mind and lead her to continue to make such attributions in the future. To give a concrete example, let us take the case of a woman, Jane, at two points in her menstrual cycle, premenstrum and ovulation. At both points she experiences tension and anxiety, as well as some adverse external circumstances. At ovulation she is likely to attribute her mood to the outer world, thus maintaining her self-image. At premenstrum, however, she is more likely to link the mood with her biological state, discounting the impact of negative external events, with more negative implications for her sense of self and feeling of control over her moods. It is likely that most women have some rudimentary sense of where they are in the menstrual cycle, due to minor physical changes, such as water retention and breast tenderness, as well as to the practical need to plan ahead for sanitary protection. In some sense, then, the menstrual cycle provides a built-in excuse for negative moods, as well as socially disapproved behaviors such as aggression and irritability, and this "excuse" then serves to perpetuate the link.

What can we conclude about the effect of the menstrual cycle on women? We can think of it as having four major influences, only one of which is the purported hormonal effect on brain activities. First of all, there is the effect of the negative and tabooed image of menstruation in the culture, which can cause long-range effects on a woman's self-image, especially during menstruation. An increase in openness about sexuality may lessen this negative effect. Second, there is the issue of physical discomfort (which is subject to wide individual differences) and the allied issue of sanitary management of the menstrual flow. The latter has undergone radical improvement in recent years with the introduction of sanitary napkins, tampons, and other devices, thus lessening anxieties and discomfort. The third influence is that of the attributional bias of women and men to assume that negative moods during the premenstrum have internal causes, and to discount possible external causes for such moods. Thus negative moods may pass unnoticed during other times of the month but be flagged as biologically caused during the premenstrum; this labelling process may then affect the woman's later attitudes so that she *expects* negative feelings in the premenstrual period and may therefore get them. Men, raised in the same culture of menstrual myths, may also label female behavior as menstrually caused, thus obviating the necessity for further analysis, even if they have no access to information as to the woman's menstrual point. Positive moods and behaviors occurring before the menstrual period, because they are out of accord with expectations and beliefs, may pass unnoticed, thus preventing disproof of the belief.

The fourth influence is the most controversial, and that is the actual hormonal input to brain activities and hence to mood and motivational states. Correlational and post hoc studies that do not attempt to delineate a biochemical mechanism fall prey to the argument that any or all of the first three factors may really be causing the positive result (if found). We have seen that the few studies that have attempted to deal with this issue (e.g., Paige's study) have come up with complex and conflicting results (i.e., attribution to other factors, such as differences in menstrual flow). Even if a link is finally established between such hormonal swings and emotional states, the hormone will remain but one factor involved in producing a particular behavior, interacting with external events, self-image, expectations and attributions, and subject to great individual variation. To take a male analogy, even though the link between androgen and aggression is reasonably well established, no one would argue that men are totally at the mercy of their hormones, or that all must act in the same way, given the similarities in their hormonal inputs. The bias in "female psychology" is to look for unitary explanations for a (nonexistent) unitary set of traits, the eternal feminine character, an attitude that has shaped scientific research no less than social expectation.

We might make a brief foray into the area of male cycles,[40] since the issue has been almost totally cast in female terms. Not only is there the supposition that male moods are independent of any (cyclical or otherwise) biological input, there is little talk of male emotionality in general, except in reference to aggression when it is ascribed as a universal human trait (though its enactors are almost exclusively male). Some attention has recently been paid to the issue of male emotionality, especially in reference to cyclical factors, but it is as yet quite preliminary in nature. The bias of confusing woman with nature and biology, perhaps because of her part in childbirth, and of ignoring male biological roots, has flavored scientific thought as well as popular conceptions.

Evidence on male cycles is at best indirect, since there is no known biological landmark against which to calibrate such fluctuations as there is in the case of the female menstrual cycle. A forty-five-year-old study documented the existence of long-term cycles in a group of industrial workers. These cycles varied from four to nine weeks in which recurrent mood states were noted. However, little work has been done since then to support or question this finding. Research on depression has correlated the rhythm of adrenal hormone secretion with mood change. A study by Halberg and Hamburger[41] in 1964 of the adrenal hormone secretion of one male indicated evidence for a monthly cycle over the sixteen years of the study. A more recent study has related the blood level of testosterone to aggression in men, but no cyclic qualities were examined.

A recent study by Doering, Brodie, Kraemer, Becker, and Hamburg[42] sought to establish a link between plasma testosterone levels and psychological states over an extended period of time (two months). No regular relationship to affective state was found, and no evidence for cyclicity either in hormonal level or affect was reported. Some relationship between plasma testosterone level and self-perceived depression was found, but the hormonal input could only account for a very small amount of the variance in psychological states. The investigators caution, however, that difficulties in assessing hormonal level through blood tests may have clouded the relationship to mood, if such existed.

Studies of circadian rhythms in humans have demonstrated the existence of many biological cycles, but the documentation of psychological ones remains to be established. Evidence for male cycles is incomplete, and the field does not seem to be receiving much research attention so that the situation in the immediate future is likely to be no better. Again, the culturally induced beliefs of researchers and the ease of landmarking the hormonal events of the cycle have facilitated research into female cycles, but inhibited that on males.

A MATERNAL INSTINCT?

To what degree does the reproductive role of the female create a constellation of psychological traits concordant with the requirements of child care? Or, to put it more directly, is there such a thing as a maternal instinct? Such an instinct would presumably act to motivate women to achieve maternity and would maintain nurturing and child-centered behaviors once maternity was achieved. Such an instinct, if it exists, might be very general in orienting women to interpersonal and emotional concerns and a nurturing role, even outside of the specific context of child care.

Animal evidence does support the concept of the maternal instinct, in that female sex hormones such as estrogen, progesterone, and prolactin seem to be implicated in the ontogeny of maternal behavior. For example, studies of ovariectomized rats given this triad of hormones demonstrated the onset of maternal behaviors, such as nest building, retrieval of pups, etc. Many studies have come up with conflicting results, due to the sensitivity and intricacy of the hormonal system and the confounding effect of exposure to the young. In many cases (including the human), early and continued exposure to the newly born animal seems to be a requirement for eliciting the caretaking behavior, regardless of the hormonal component.

Treatment of virgin and ovariectomized females with testosterone (the male hormone) has been related to pup killing. Castration of male rats, on the other hand, seems to increase the likelihood of caretaking behavior. Prior experience with the young seems to be a potent determinant of maternal response as well. Any full interpretation of animal maternal behavior would have to include consideration of experiential factors as well as hormonal ones. Present evidence seems to imply that hormonal factors operate in concert with the eliciting properties of the young pup so that its presence acts to maintain such behaviors. Studies of maternal behaviors in animals have concentrated on such species as rabbits, mice, and doves. A reviewer has concluded, "There is little known about the role of hormones in initiating and maintaining maternal behaviors in humans. The suggestion that they are important would be, at best, speculative."[43]

Let us be speculative, however. If such an instinct were to exist, how would we recognize it? Its results would presumably be a greater female desire for children, involvement in child care, and orientation to the affective interpersonal world (concordant with the emotional demands of children). We clearly have a world in which all of the above are present, yet the nature of the family system and the structure of sex roles and societal participation would be more than enough to maintain the status quo without the intervention of hormonal factors. For example, the frequently found early sex differences in person vs. thing orientation[44] could be explained by parental shaping and child identification and experience with sex-typed toys (such as dolls). But where do the dolls come from? Do they come from the maternal instinct or from societal conditioning? We come again to the insoluble dilemma of the interlocking nature of biological and cultural factors. Clearly the biological demands of the reproductive cycle make a maternal instinct plausible but so do the societal demands of the family system.

Paternal behavior[45] is a rarity in the animal kingdom, although it does exist to some degree in some primate species. Nowhere does the primary responsibility for care of the young belong to the males, however, and in many primate species, little participation by males is noted. Studies of mice have indicated that paternal behavior can be increased by exposure to the young. Male mice retrieved the young, groomed them, built nests, and even assumed a nursing posture when exposed to a litter of young pups, just as virgin females did in the same circumstances, thus suggesting that the role of hormonal factors was minimal. However, results vary quite a bit by species (e.g., for rats the role of hormones seems more important than in mice), so that generalizations are difficult to make.

In the human case, the incidence of postpartum depression, often involving outright rejection of the infant, and of child abuse would seem to argue against the sanctity of any notion of a maternal instinct operating independent of other psychological and sociological factors. There is considerable variation in interest in child rearing among women, though on the average it seems always to exceed that of men at all ages. Girls who have the adrenogenital syndrome (with an excess of prenatal androgen) do seem to show less interest in infants than normal girls so that the inhibiting role of androgen may be a factor. The aggressive, activity-instigating quality of androgen may not be a direct antagonist to maternal behaviors, it may simply make those behaviors less likely by introducing competing responses, such as aggression and high physical activity. The sustained attention and orientation toward other persons needed for child care may be interfered with by the instigating qualities of androgen.

The socially stigmatized state of the "barren" (childless) woman may be a better explanation of the motivation to have children than the maternal instinct. In many societies, child bearing and child rearing simply provide the only socially acceptable adult roles for women. When this is no longer the case, and when birth control has made other alternatives possible, the birth rate drops and the number of childless (or, more positively, childfree) women rises. Recently a national organization of nonparents has arisen to justify its state and propagandize against what it calls "pronatalism, the myth of mom and apple pie,"[46] or, more prosaically, the cultural push toward parenthood. A theory of maternal instinct could be compatible with such emotions, granting individual variation in instinctual strength and the overriding contribution of social psychological factors. But we are left with a rather weak version of the instinct, one that can be readily shaped by other factors—surely not the image of the overriding maternal instinct we began with.

In conclusion, then we really have no conclusion, but rather a promise of a continuation in our chapter on the family. Animal evidence does implicate hormonal factors but it also gives support to the eliciting quality of the young and the experiential context. Direct human evidence is missing, but we might accept some hormonal undercurrent as a concession to adaptive plausibility, accompanied with a heavy dose of societal conditioning and patterning of familial roles.

We turn now to the case of psychosexual abnormalities, which hold out the tantalizing hope of the separation of biological and social factors. However, given the complexities of the research and theoretical issues involved, no quick answer appears at hand. The area of research is important, however, for its raising of the issues and the promise of better solutions in the future.

PSYCHOSEXUAL ABNORMALITIES

Psychosexual abnormalities can be considered to be "natural experiments," permitting us to see the operation of sex hormones in comparison with the normal case. In recent years a number of these abnormalities have been systematically studied and the results of such studies related to the area of sex role differentiation. A leading researcher in this field has been John Money[47] of Johns Hopkins University, and we shall report on his work at several points. An exhaustive study of such cases is beyond the scope of the book, however. Instead, we have chosen three areas on which to concentrate our attention: the adrenogenital syndrome, matched pairs of hermaphrodites, and transsexualism. These areas shed considerable light on the nature of sex role development, affording some glimpse of the relative weight of biological and social factors. The reader is advised to review the explanation of prenatal sexual differentiation offered earlier in the chapter as a prelude to this section.

ADRENOGENITAL SYNDROME

Females with the adrenogenital syndrome are a "natural experiment" testing the effects of prenatal androgen on subsequent behavior and gender identity. This group has received much scientific attention because the syndrome closely parallels the condition of fetal animals given androgens experimentally. These experiments have shown that androgens produce definite masculinizing effects in sexual and aggressive behavior. Individuals who exhibit the adrenogenital syndrome have defective adrenal glands which release excessive amounts of androgen instead of cortisol during the critical period of sexual differentiation before birth. The adrenogenital syndrome (AGS) is genetically transmitted as a recessive trait. Genetic males with AGS usually do not have abnormal genitalia but need cortisone treatment to prevent early puberty (caused by the oversupply of androgen) and to correct other possible side effects of the lack of cortisol. This group is not usually of special interest to sex role researchers since the syndrome almost never has any effect on gender identity in genetic males.

The group that is of prime interest, however, is that of genetic females (XX) with AGS. Because of the medical side effects of lack of cortisol (salt loss, hypertension), the syndrome is detected early in many of these girls, and they are put on cortisone replacement therapy so that postnatal masculinization is prevented. However, those afflicted with this condition have received a heavy dose of prenatal androgen and its enduring effects, if any, are the subject of study for sex researchers. An early study by John Money on fifteen girls with AGS turned up a set of provocative and controversial findings. The girls in the AGS group (from four to six years old) were found to show a higher level of tomboyish behavior, including more rough play, less interest in dolls, and a greater preference for boys than girls as playmates than girls in a control group. Another study by Money and Lewis[48] on seventy AGS individuals pointed up a higher IQ than would be expected in a normal population (a mean of 109.9 as opposed to the expected mean of 100.0). Studies of individuals with other endocrine abnormalities have not turned up this effect (in fact, several are associated with reduced intellectual potential). Another group of children, born to mothers who took progestinic drugs prenatally, also show an elevated IQ. Progestinic drugs often have a masculinizing effect on female genitalia at birth, although there is no known increase in circulating androgen.

The provocative findings on the girls with AGS prompted other researchers to do further studies on the phenomenon. Ehrhardt and Baker[49] conducted a study in which AGS patients were compared with their normal siblings on behavioral measures. The point in this comparison was to control, insofar as possible, for the contaminating effects of social environment on behavioral dispositions. Thus, if tomboyish behavior were found to be above average in families of AGS girls, the exclusive contribution of the hormonal input could be ruled out in favor of a less mysterious explanation of support from the social environment as manifested in the socialization of all children in the family. If such behaviors were found to characterize AGS children and not their normal siblings, the hormonal explanation would receive support. In this study, boys with AGS were studied as well, on the supposition that additional prenatal androgen might stimulate supermasculine behaviors even among boys, over and above that seen in their unaffected brothers. The children studied ranged in age from four to twenty-six years, with most at mid-childhood and early adolescence, and all were receiving treatment for AGS, usually from early childhood. The comparison of AGS girls with their normal sisters indicated that they were more often described by others as having high energy expenditure and that they preferred boys to girls as playmates. Tomboyish behavior and low interest in dolls, infant care, and marriage also characterized this group. For example, babysitting was described as being significantly less frequent among AGS girls than the normal sample. Interest in jewelry, makeup, or frilly clothes was also lower in this group. There was a nonsignificant trend for AGS girls to be described as initiating fights more often than their normal sisters. It is important to note here that many of these behaviors are reports from parents, siblings and the AGS girls themselves, and do not represent direct observation. Given the masculinizing effect on the external genitalia, it is possible that this changed the family's view of the girl (as well as her view of herself) so that male sex-typed behaviors became more likely and female sex-typed behaviors became less likely. However, the findings also suggest that some hormonal priming mechanism may be at work.

In the same study, male AGS patients were compared with their male siblings to detect any "excess" of male sex-typed behaviors. This comparison is likely to be less contaminated by parents' and patients' expectations, since in this case the hormonal effects are concordant with the child's genetic sex rather than discordant with it as in the female AGS case. Interest in sports and physical activities was rated higher for AGS boys, but otherwise there was no difference between the samples in initiation of fighting, choice of playmates, tomboyish behavior, or feminine-typed interests. Thus the effect on behaviors seems to be less dramatic here than in the female case, possibly because of hormonal factors, familial treatment, or both. That is, the additional androgen may have less effect on males than females because of the already existing amount of circulating hormone in males; the excess may have little additional impact. It is also possible that families of AGS boys were less disrupted by the AGS syndrome than families of AGS girls (because of the concordance of the effects of the condition with genetic sex) and hence did not transmit any special expectations to the boys, whereas this might have been the case for the AGS girls. It should be emphasized, however, that even in the case of the girls, all were clearly identified with the female role and seemed to be adopting a normal heterosexual orientation. According to the authors, "they presented an acceptable pattern of tomboyish behavior in this society, not unlike tomboyism in normal females except that it occurred significantly more often in the AGS sample than in either the sibling or the mother sample."

Another study by Baker and Ehrhardt[50] on the AGS syndrome aimed to clear up the furor caused by the earlier Money and Lewis study on IQ, which seemed to suggest that prenatally circulating androgens increased intellectual abilities. The Baker and Ehrhardt study compared AGS patients with their siblings and parents and found no significant differences in IQ, although the scores as a whole were higher than would be expected in a random distribution of subjects. In other words, the Money and Lewis results seemed to be explained by the fact that for some unknown reason virtually all members of AGS families had high IQs, not just the AGS children themselves. Thus it is unlikely that prenatally circulating androgen causes the effect since family members without this defect also show the rise. Two explanations are offered by Baker and Ehrhardt to explain the results. One is that the recessive AGS gene, which is carried by most members of AGS families, is linked in some way to a genetic predisposition to higher intellectual abilities. The other, more pedestrian explanation, is that there is some unknown selection factor that makes it more likely that high-IQ AGS families will come to the attention of hospital detection clinics for endocrine disorders. Two facts make this explanation unlikely. One is the report that patients with other endocrine disorders do not show this IQ increase, and the other is the full spread of social class among patients and parents, and the lack of relationship within this group between SES and IQ. Therefore, the effect remains something of a mystery, but perhaps not one that is of special interest to sex role researchers since the link between androgen and IQ does not seem to hold up on closer scrutiny.

MATCHED PAIRS OF HERMAPHRODITES

An especially instructive case of the impact of biological and social factors can be given in the instances of persons with identical physical anomalies who are given different upbringings with regard to assigned sex roles. John Money[51] has studied a number of such cases and we shall report on one of them. Both children were in fact genetic females with the previously discussed adrenogenital syndrome, which caused the genitals to appear masculine at birth. Child A was thought to be a boy at birth because of genital appearance but was soon correctly identified because of other physical manifestations of the syndrome. The child was reassigned as a girl, and the parents accepted this and were able to explain the situation to family and friends. Thus it is likely that the child was brought up with a minimum of confusion as to her sexual identity. The child had tomboyish interests and later showed more interest in career plans than early marriage. Surgical correction of the masculine-appearing genitals was accomplished at the age of two, and breast development and menses occurred during adolescence (as would be expected in a normal genetic female). Except for her tomboyish tendencies, Child A was a typical female, in accord with her genetic and social identity.

Child B in this matched pair was also a genetic female with adrenogenital syndrome but was misidentified at birth as a male with a small phallus and undescended testes. Surgery to masculinize the child's genitals was attempted unsuccessfully. Finally, at three-and-a-half, the child was correctly diagnosed as being a genetic female with AGS. The child had been raised as a boy for its entire life and the decision was made to maintain the masculine gender identity by the administration of androgen therapy and surgery. The child manifested a masculine identity, associating with other boys and sexually orienting himself to women. Some difficulties might be expected in later life, however, due to the undersized penis and prosthetic testes, but the boy's view of himself is in accord with his social identity rather than his biological one.

Money reports the case of another child, a genetic female with AGS who was thought until age twelve to be male and was raised that way. In adolescence he developed attraction toward women and thought of himself as male in every way despite his growing breasts and female hormonal production. Male sex assignment was maintained and aided by androgen replacement therapy after the correct diagnosis had been made. Money notes that in most cases the socially imposed identity is the accepted one as long as there is no ambivalence and uncertainty about the child's sexual identity. That is, difficulties seem to arise only when there is parental confusion about the child's sex role identity and the child is raised inconsistently, or when an abrupt change is imposed after an identity has already been established. A "critical period" between eighteen months and three years has been hypothesized as marking the time after which sexual reassignment is unwise because sex role identity has already been established. Research by Money and the Hampsons has tended to support the idea of *psychosexual neutrality* at birth with the major determining input coming from environmental factors. Others (e.g., Diamond, Stoller[52]) oppose this view and lay heavier emphasis on biological input, citing case studies to support their view. These cases usually involve some uncertainty about sex assignment and inconsistent socialization, so they do not constitute final proof of this view. Of course, in all normal cases the child's genital appearance at birth *is* concordant with hormonal factors (both prenatally and postnatally) and is the basis for sexual identification and later sex role identity.

One of Money's most dramatic matched cases is that of a normal set of male identical twins in which sex reassignment for one was necessitated by the accidental surgical loss of the penis during circumcision at the age of seven months. Since the boy now lacked a penis, it was decided to reassign him as a girl with a change in name, clothing and hair style at seventeen months. Surgical creation of a vagina was begun and female hormone therapy was planned for a later age. Money has followed these children up to age seven and reports that the reassigned twin is differentiating normally as a female, displaying typical feminine interests in clothing, toys, and domestic activities. The mother did report tomboyish behavior but indicated that she was attempting (with success) to deemphasize these actions and reinforce more feminine behaviors.

Research on sexual anomalies such as AGS seems to support the idea of the overriding effect of sex of rearing on the person's sex role identity. This finding does not necessarily imply that the person is totally plastic at birth with regard to sexual orientation, only that contrary socialization can generally overcome the hormonally imposed direction if begun early and consistently enough and if supported by the person's environment and view of himself or herself. Diamond has put it succinctly:

> Sexual behavior of an individual, and thus gender role, are not neutral and without initial direction at birth. Nevertheless, sexual predisposition is only a potentiality setting limits to a pattern that is greatly modifiable by ontogenetic experiences. Life experiences most likely act to differentiate and direct a flexible sexual disposition and to mold the prenatal organization until an environmentally, socially and culturally acceptable gender role is formulated and established.[53]

But what of the cases of hormonally normal individuals raised in one sex and living that sex even until well into adulthood and then deciding on sex reassignment because of a compelling desire to live as the other sex? What can these cases tell us about the origin of sex role identity and its relationship to biological factors? For some preliminary answers to these questions, we turn now to the case of transsexualism.

TRANSSEXUALISM

It is important to distinguish transsexualism from two other conditions, homosexuality and transvestism, with which it is often confused. Both homosexuals and transvestites maintain their own sex role identity concordant with that of their birth and upbringing. The only difference is their sexual orientation to the same sex (in the case of homosexuality) or their predilection for dressing as the opposite sex (in the case of transvestism). In pure cases of homosexuality or transvestism, there is no doubt in the person's mind about his or her sexual identity; that identity has been maintained intact since birth. Transsexualism, on the other hand, is marked by a fundamental feeling on the part of the person that he or she was born into the wrong sex and should be of the other sex. This feeling often goes so far as to become a desire for genital surgery. The transsexual generally manifests aspects of both homosexuality and transvestism. That is, from an external perspective, the person is sexually attracted to his or her "own" sex and wishes to dress as the "opposite" sex. But from an internal point of view, the transsexual's sexual orientation and dressing preference are entirely consistent with his or her "right" sex, which happens to be different from that of his or her genitalia and socialization. The organizing principle in the transsexual's perceptions is the gender identity of the other sex; this determines all other aspects of life.

A glimpse into the life of a transsexual was given by Jan Morris in her recent book, *Conundrum.*[54] Born and raised male, as James Morris, she describes herself as having the conviction of being a girl from an early age. Despite this, she lived a conventional male life, serving in the army, marrying, and fathering children. Throughout her life, she recalls a distaste for masculine activities and an attraction to feminine ones. Sex was a relatively unimportant part of her life, but after homosexual and heterosexual experiences she laments "though my body often yearned to give, to yield, to open itself, the machine was wrong. It was made for another function, and I felt myself to be wrongly equipped." For Morris, though, the fundamental mistake seems not to be a sexual one in the strict sense of the word, but one of gender identity. She notes:

> To me gender is not physical at all, but is altogether insubstantial. It is soul, perhaps, it is talent, it is taste, it is environment, it is how one feels, it is light and shade, it is inner music, it is a spring in one's step or an exchange of glances, it is more truly

> life and love than any combination of genitals, ovaries and hormones. It is the essentialness of oneself, the psyche, the fragment of unity. Male and female are sex, masculine and feminine are gender, and though the conceptions obviously overlap, they are far from synonymous.[55]

Many other transsexuals report sexual activities as a relatively small ingredient in their gender identity, although they are drawn to individuals who complement their assumed identity and hence often wish a surgical change to facilitate the relationship, or simply to remove a discordant element in their new identity. In the case of Jan Morris, she began cross-dressing and living as a woman in adulthood and at age forty-six underwent a transsexual operation for removal of the penis and its replacement with a vagina. She reports a close friendship with her former wife and a comfortable "aunt" relationship with her children. The transsexual transformation was a welcome one for Jan Morris, for now her external social identity and her internal identity could be one.

Inquiry into the background of transsexuals[56] usually reveals a very early desire to be of the other sex. Male transsexuals often recall a childhood of female preferences in the choice of toys, clothing, and playmates, as well as a female self-concept. Oftentimes parents of these feminine-acting boys were reported as being overtly or covertly supportive of their sons' actions, often supplying makeup, dresses, etc. In some cases, such a boy's physical beauty inclines adults around him to treat him in feminine ways. Some investigators also think that the lack of a strong male identification figure and a controlling, dominant mother are other contributing factors, but this has not been established in all cases. Any familial theory would also have to take account of the fact that most transsexuals have perfectly normal sex-typed siblings who presumably grew up in the same home atmosphere, although, of course, variations in treatment of different children in the same family are surely present. The search for a hormonal determinant of transsexualism has so far proven futile, with testicular defects accounting for only three out of the multitude of male cases reported in the literature. An environmental cause seems indicated, although it is possible that errors in prenatal sex-typing may be present that are undetectable by present methods.

The case of the female transsexual is considerably rarer, accounting for only one-fourth to one-seventh (depending on estimates) of all transsexuals. Explanations for this disparity include the practical one of the greater difficulty of creating a penis than a vagina, which may cause fewer women to appear for surgical treatment (where most transsexuals first come to the attention of physicians). In addition, women acting in a masculine manner may be subject to less social sanction than men acting in a feminine manner, so that more female-to-male transsexuals may be able to manage their lives without coming to medical attention. A more interesting hypothesis argues that there are more male transsexuals because of the greater possibility for error in the endocrine creation of the male through the prenatal addition of androgen. Thus far we do not have a final answer to this question, but inquiry into the background of female transsexuals produces a picture similar to that of their male counterparts. An early history of tomboyish behavior, clothing, and playmate preference often is evident, sometimes in the context of a family with a dominating father and an inadequate mother who does not provide an identification figure for the girl. Tomboyish behavior usually does not arouse as much concern among parents as does feminine behavior in boys, so that this behavior is not always made an issue of to the child.

The typical case of the transsexual is that of an individual who is extremely resistant to therapeutic efforts aimed at making his self-perceptions and behaviors consonant with his biological sex. David Barlow[57] and his associates have reported on a case of a male transsexual who was successfully shaped by behavior therapy to give up his female characteristics and to assume a male identity, but this is clearly an exceptional case. The nature of transsexualism remains something of a mystery, with possible biological inputs unknown at this time, but with somewhat suggestive findings about the determining role of the child's early social environment.

The study of psychosexual abnormalities, then, can reveal some interesting findings relevant to sex role concerns, but it has not yet given us indisputable evidence as to the relative weighting of biological and environmental factors. This literature suggests that both factors are significant and that biology is malleable to a surprising degree when faced with the impact of socialization.

A CLOSING NOTE ON THE BIOLOGICAL MAINTENANCE SYSTEM

Instead of providing clear-cut answers, our inquiry into the biological roots of sex roles has raised more psychological and sociological questions. At every point in our discussion of biological inputs, we have had to consider the simultaneous contribution of socialization and societal factors. By now the reader may be willing to accept the idea that biology, psychology, and sociology are analytic categories of convenience, not inseparable, uncontaminated inputs. What does it mean to say that biological factors are more "basic" or uncheangeable than cultural ones? Such statements can only take on meaning if there is a clear way of distinguishing biological inputs from cultural ones in the life of a person or of a social structure. Given our current state of knowledge about the role of hormones in human behavior, we are simply not at the stage where such mechanisms can be specified with any degree of certainty.

Based on the current rudimentary evidence, it seems reasonable to speculate that sex hormones may have an early differentiating function in terms of propensity to aggression, which may lay the groundwork for other more complex sex role behaviors if environmental conditions are favorable. Another major contribution of biology probably comes in the nature of the reproductive system itself, in the simple fact that it is women who conceive and produce children. The complexity of the family system has been overlaid on this "simple" fact, leading to wide disparities in the life plans of men and women. Much of the socialization experience of the child is preparation for participation in this system, as we shall see in the next chapter.

SUGGESTED READINGS

Jesse Bernard. *The Future of Motherhood*. Baltimore: Penguin Books, 1974 (p). Scholarly treatment, emphasizing sociological and economic factors. Very readable.

Jayme Curley, et al. *The Balancing Act: A Career and a Baby: Five Women Describe the Problems and Rewards*. Chicago: Swallow Press, Chicago Review Press, 1976 (p). First-person accounts of motherhood by five career women.

Janice Delaney, Mary Jane Lupton, and Emily Toth. *The Curse: A Cultural History of Menstruation*. New York: Dutton, 1976. Lively, readable account of how (and perhaps why) menstruation is perceived as such a special event in most cultures.

Richard C. Friedman, Ralph M. Richart, and Raymond L. Vande Wiele, eds. *Sex Differences in Behavior*. New York: John Wiley & Sons, 1974. Collection of articles reporting recent research on biological factors in sex roles such as hormonal effects. Some coverage of psychological areas like early infant sex differences as well.

Richard Green. *Sexual Identity Conflict in Children and Adults*. Baltimore: Penguin Books, 1975 (p). Account of research on children whose gender identity does not agree with their biological sex. Accent on psychosocial factors. Green is a leading researcher in the area.

Shere Hite. *The Hite Report: A Nationwide Study of Female Sexuality*. New York: Macmillan Co., 1976. Well-received report based on a survey of 3,000 women, ages fourteen to seventy-eight, on how they perceive their sexuality.

Jane Lazarre. *The Mother Knot*. New York: McGraw-Hill, 1976. Personal, compelling account of how one woman approached the present-day conflicts and rewards of maternity.

John Money and Anke A. Ehrhardt. *Man and Woman, Boy and Girl*. New York: New American Library, 1974 (p). Well-done presentation of findings in the area of psychosexual abnormalities by a team that pioneered research in the area.

Jan Morris. *Conundrum*. New York: New American Library, 1975 (p). First-person account of a transsexual's life and sex change.

Adrienne Rich. *Of Woman Born: Motherhood as Experience and Institution*. New York: Norton, 1976. Poet's report of her own experiences as well as some sociohistorical material. Feminist perspective taken on motherhood as a societal institution.

Klaus R. Scherer, Ronald P. Abeles, and Claude S. Fischer. *Human Aggression and Conflict: Interdisciplinary Perspectives*. Englewood Cliffs,

N.J.: Prentice-Hall, 1975 (p). Excellent treatment of aggression from a variety of perspectives.

Mary Jane Sherfey. *The Nature and Evolution of Female Sexuality*. New York: Random House, 1972 (p). A controversial theory of societal repression of female sexuality, sexuality that the author contends is greater than that of men.

C. A. Tripp. *The Homosexual Matrix*. New York: New American Library, 1976 (p). Comprehensive, readable review of the subject.

Paula Weideger. *Menstruation and Menopause: The Physiology and Psychology, the Myth and the Reality*. New York: Knopf, 1976. The role hormones play in women's lives and what society makes of this.

2 The Psychological Maintenance System

Sex role identity is an overdetermined fact of human development. So many factors conspire to make it happen that it is a hopeless task to try to sort them out and single out one or two as unitary causative agents. The fact that so many seemingly conflicting theories of sex role socialization have come up with positive results supports the notion that sex role identity is being impressed on the child from a multitude of directions. It makes sense to try to enumerate some of these directions and to indicate the relative weightings of each, insofar as that is possible.

One might think of sex role socialization as coming in two parts for the child: a paradigm of action for the present and a plan of life for the future. Both parts are complementary and function as a whole, but for the child there is a distinction. Not only is the child shaped for the present; he or she is also prepared for the future, and the shape of the future influences the course of the present. Thus, to take the simplest case, the selection of toys is premised on different adult life plans: trucks and soldiers are chosen for boys, dolls for girls. At the same time, these toys mark out different childhood roles for the two sexes: action and physical activity for boys, sedentary interests for girls.

It is perhaps easiest to organize the factors involved in sex role socialization around the primary agents of socialization: parents, schools, peers, and symbolic agents. However, before considering these agents, it is necessary to touch upon the issue of biological predispositions. The argument goes somewhat as follows: socialization could not be imposed upon a population of recalcitrant children; the demands imposed must be in accord with the different biological roots of behavior in males and females. Therefore to understand sex role socialization, we must first understand this physiological base. In the first chapter we discussed this issue at some length, considering possible biological inputs throughout the life cycle. We saw that this research is in a very preliminary stage, hampered by the fact that we do not yet fully understand the functioning of the brain and hormones in behavior. Evidence from the study of psychosexual anomalies may be suggestive as to the course of normal development, but even this evidence is somewhat fragmentary and often contradictory. We will consider studies of very early sex differences in the section on parents as agents of socialization, since at the newborn period it is the parents, particularly the mother, who respond to these differences (if they exist) and create the very specific child-parent system of behaviors and responses that will lay the groundwork for later sex role socialization. Michael Lewis, a well-known researcher in the area of early sex differences, notes: "Our research experience indicates that as early as we can find individual differences as a function of the sexual dimorphism in the infant so we can find a society expressing differential behavior toward the two sexes."[1] Biological predisposition, then, is simply not clearly enough established as a determining factor apart from parental input to justify considering it as the first step in sex role socialization. It probably *is* a step, but not necessarily the primary one; in any case, biological input is so intimately intertwined with the action of socialization agents and the structure of society that to consider it apart from these factors would be only an act of analytic separation.

We turn now to a consideration of the primary agents of childhood sex role socialization and shall begin with parents, probably the most important of all agents. Because of the continuity of the parent-child bond over many years, the power of the parent over the child, and the (usually conventional) sex role identities of the parents themselves, parents are in a unique position to shape the course of their child's life. It should be emphasized, however, that parents do not act alone, and the contributions of other agents can work to reinforce or oppose their demands. We shall therefore also consider the effects of the school environment, peers, and symbolic agents of socialization. Finally, we shall turn to the "results" of socialization, to what is known about modal sex differences in cognitive abilities and personality traits.

SEX ROLE SOCIALIZATION

PARENTAL AGENTS

Two major mechanisms can be seen as being primarily responsible for parental sex role socialization: differential treatment and identification. Both of these processes are at work in the family and in the larger socialization context. That is, the child will be dealt with differently by parents and other adults and institutions as a function of sex and will model himself or herself according to different standards, depending on sex. These two processes are interdependent and tend to reinforce each other. Probably the larger share of socialization is taken over by the process of identification, as it is a theoretically more parsimonious concept than is differential treatment. Thus, it is unlikely that in each person's life enough instances of sex role relevant behavior are manifested and consistently reinforced in ways that support sex roles for this process to be the more important. The hit-or-miss quality of differential treatment is simply insufficient to produce the development of role identity by itself, although such treatment presumably contributes its share. Identification, on the other hand, provides a cognitive-emotive bridge between the person and the perceived standards of the social structure. Once a person becomes linked with the same-sex model in the family and the more abstract standard in the social structure, an entire repertory of social reactions and behaviors is made available, beyond the small subsample any given individual is likely to experience directly.

But none will deny that definite sex-typed expectations do exist and are transmitted in direct and subtle ways. For example, the often-repeated example of parents' suppressing male emotionality (and, in particular, crying) and perhaps encouraging female emotionality, no doubt exists, and certainly scenes are enacted in which boys are explicitly told that "boys do not cry" and are pressed to conform to this standard. A far more potent shaper however, may be the child's identification with the same-sex parent (who is seen as responding in an emotional or nonemotional way), as well as the plethora of societal and media models available. The child may thus come to anticipate the reaction of others to his or her emotional outbursts and moderate them according to sex role standards and situation.

We will turn now to a consideration of these two complementary processes—differential treatment and identification—and let the reader make up his or her own mind as to the relative contribution of each. Perhaps some effort at recall of one's own socialization experiences might be helpful, although with the caveat that such recall is likely to be selective. In other words, specific instances of explicit tutoring in the appropriate behavior may be more memorable than the more subtle and perhaps more pervasive influence of shaping oneself to conform to familial and societal models.

Differential Treatment

Several methodological difficulties underlie any inquiry into differential treatment. First of all, the choice of the domain of behavior is critical, as is the method of observation. It is more or less generally agreed that men and women are socialized differently, but this difference often is difficult to identify under close scrutiny of specific tasks, especially in laboratory situations. The laboratory context may elicit atypical responses from both children and parents, and the specific task may not be a good one for demonstrating sex differences (or anything else, for that matter). The size of the behavioral unit under observation may simply be too small to show the expected effects. In this context, a very revealing study[2] was recently published that showed pervasive differences in the furnishing of boys' and girls' rooms, differences that were presumably a good index of differential parental attitudes and expectations. The study was done in a middle-class area of a university community, a locale that would presumably be on the less differentiated end of the sex role socialization spectrum. Yet, "the rooms of boys contained more animal furnishings, more educational art materials, more spatial-temporal toys, more sports equipment and more toy animals. The rooms of girls contained more dolls, more floral furnishings and more 'ruffles.'"[3] More specifically, the 48 boys' rooms had 375 vehicles, the 48 girls' rooms had 17; 26 of the 48 girls had baby dolls, only 3 of the 48 boys had such dolls. In this very global measure, clear differences emerged, but the socialization studies often look at fragmented areas which may hide the general pattern of differential socialization.

Another methodological issue in this area concerns the behavior of the child as a trigger of parental response; rather than viewing the parent as the instigator of the child's behavior, one might label the parent's behavior as a function of the child's disposition and responses. The issue becomes more complicated if we consider the child's sex as influencing and interacting with parental responses. Thus, the observation that parents tend to talk more to female children might be seen as a response to the female child's greater receptivity to social inputs rather than to any concerted parental "plan" to shape the child into a good little girl. One thing is fairly certain: only a small part of a parent's behavior is consciously directed toward molding a specific sort of child. Most child-directed behavior is a complex combination of the child's responses and the parents' image of what that child *is,* a large part of which may have to do with the sex role standards of the culture and the parents' own sex role concepts. Often these concepts are quite deeply ingrained in the parent (as in any "well-socialized" person) and may be at variance with stated ideological beliefs. For example, interviews[4] with avowedly feminist mothers revealed that boys were often reared in stereotyped ways despite ideological commitment to the contrary position (somewhat greater "success" was reported in the "shaping" of girls' behavior, perhaps as a function of the low evaluation of female traits and the high evaluation of masculine traits, e.g., aggression). The biographies of career women often show a professionally-oriented mother, who could be effective both as a "shaper" and model for the child. One suspects that the mother's role as model would be the more potent determinant.

To prevent feedback from the child from affecting parental response, a number of studies have adopted the innovative, albeit somewhat artificial, method of labeling the same stimulus as either male or female and seeing if the label alone can lead to a differential response. The separation of parental reaction from child behavior is of course an artificial one; in real life, the child and parent develop a complex feedback network in which each influences the other's response. An often-cited study in the literature[5] featured a child's voice that was alternately presented as either that of a four-year-old girl or boy to parents of nursery school children. A rather interesting cross-sex effect emerged: fathers were more permissive in both aggression and dependency situations to girls, while mothers acted in a similar way toward boys. There was no overall effect for a boy's or girl's voice in eliciting sex-typed responses; the sex of the parent was an important mediating variable. A similar study[6] involved the alternate labeling of a neutrally attired baby girl as male, female, or with sex not specified. Men and women then interacted with the child, and measurements of different orientations toward it, depending on sex labeling, were taken. No striking differences were found, except that more dolls were offered as toys to the baby "girl" than to the "boy." With the child whose sex was not identified, women showed more contact with it than men. All subjects were confused by the fact that the child's sex was not revealed, as one of their first inquiries was to ask the sex of the child.

A more recent study[7] in this genre used a videotape of a three-year-old boy performing various activities (telling a story, coloring pictures, putting together a puzzle, doing a somersault) and talking to the experimenter. The child was chosen because he was not readily identifiable as to sex and had an ambiguous name, Chris. Half the undergraduate observers were told Chris was a boy, the other half that he was a girl. Ratings of his behaviors were then solicited from the subjects. Although a few ratings were marginally different for the boy label as opposed to the girl (mischievous, lovable, loud, energetic, extroverted, potential for intellectual achievement), the biggest effect appeared when the subject's own sex was considered as a variable in the ratings. As in the previously mentioned study, cross-sex effects were most significant. Women were more positive in their evaluations if they thought the child was a boy, and men were more positive when the child was identified as a girl. This bias extended to predicting better performance following success and higher ability following failure in the child observed in the cross-sex case. Several explanations have been offered for this effect, notably by Rothbart and Maccoby, the authors of the previously mentioned study, and Gurwitz and Dodge, the investigators in the study described here. Clinical explanations favor the idea that the parent may feel a sense of rivalry with the same-sex child or may be reminded of his or her own negative behaviors, thus being moved to downgrade the same-sex child. A Freudian-tinged explanation would advance the notion of a closer bond between opposite-sex parent-child pairs because of the sexual nature of the association. In the Gurwitz and Dodge study, however, subjects were undergraduate students, not parents, so one would have to generalize the effect to all adult-child interactions.

But, of course, in the "real world," the behaviors of the child and of the parent form an intrinsic system, including as an important component the parent's knowledge of and expectations about sex roles. One might begin an inquiry into the origins of sex roles by asking what is known about differing behavioral predispositions of male and female babies.[8] A few differences have been stably demonstrated in several studies, including greater excitability and fussing in male infants, and greater oral sensitivity in female infants. The first finding becomes rather complicated to interpret when we note that most male infants are circumcised, surely a possible stimulus to early excitability. There also seems to be a tendency for parents to begin differential sex role socialization at a very early age, thus obscuring the nature-nurture issue. For example, it was found that newborn girls are spoken to and smiled at more than boys at feedings, and that boys are touched more than girls at this very early stage. It is hard to know whether these differences have been preceded by real differences in the infants themselves or arise through the transmission of sex role expectancies. For example, in the case of the higher frequency of smiling and talking directed at girls, we have the possibility of an act brought about by several factors. That is, the baby girl could be evidencing more spontaneous mouthing behavior, as has been found in previous studies,[9] and this could be acting as a stimulus for parental imitation. In addition, however, the parents could have sex-stereotyped notions of the greater "sweetness" of baby girls and be behaving accordingly. Other research does show that parents seem to be more concerned with the social behavior of girls than with boys. For example, in one study of girls parents spent more time in getting their baby to smile and talk when told to do so than did parents of boys. In terms of the structure of sex roles, parents' greater interest in the expressive behavior of female infants fits into the female sex role standard, and they could be responding to their knowledge of adult sex differences and shaping the child's behavior accordingly. It is more reasonable to suppose, though, that at least some of the variance is accounted for by the child's differential responsiveness.

Three studies of early parent-infant interaction bear looking into; two were done with humans, one with monkeys. The first, by Moss,[10] showed a tendency for more interaction with, more holding and stimulation of, and more looking at male infants. Interpretation of these findings is clouded by the greater fussiness of male infants as compared to female. When fussing and crying were statistically controlled, most of the sex differences disappeared, except for more maternal stimulation of males and more maternal imitation of females, two effects that were strengthened by the statistical manipulation. Children in this study were observed at three weeks and three months of age; the effects stated above were stronger for the earlier age period. Infant fussiness at three months was responded to differently as a function of sex: female infants were spoken to, imitated, and looked at; male infants were given close physical contact.

The second study by Lewis and Weinraub[11] confirms the above findings: "In the first few months of life boys receive more proximal stimulation, such as rocking and handling, and girls receive more distal stimulation, such as talking and looking from their mothers."[12] But boys also seemed to be "weaned" from proximal stimulation at an earlier age than girls, with the result that girls receive more distal contact at all ages and more proximal contact than boys after three months of age. The third study, of macaque monkeys,[13] revealed that males move away more rapidly from close contact with the mother than do females, and that females seemed to differentiate the mother from other female adults more readily at an earlier age than did males. Some evidence is thus available to suggest a genetic predisposition for proximity behavior and attachment in females, with the opposite occurring for males.

The issue of differential socialization, then, is exceedingly complex, complicated as it is by possible sex differences in child behaviors interacting with parental expectations and behavioral styles. The parent-child dyad must be looked at as a system, not just a one-way communication from parent *to* child.[14] Given these caveats, we might be surprised to note that laboratory studies of parental behaviors "revealed a remarkable degree of uniformity in the socialization of the two sexes,"[15] according to a recent review of the literature by Maccoby and Jacklin. As we noted earlier, however, there is abundant evidence of differential socialization outside of the laboratory, e.g., in the survey of toys in the home mentioned earlier. Parents generally have highly codified views of the behaviors appropriate for each sex and are not reluctant to share these with the child, both overtly through verbal instruction and covertly through provision of toys and encouragement of activities. Parental sex-typing practices have two components: encouragement of the appropriate behaviors and discouragement of the inappropriate ones. In the case of the latter, it is the deviant boy who most often comes in for criticism. Therefore, so-called "sissy" behavior is much more heavily sanctioned than "tomboyish" behavior. Maccoby and Jacklin's interpretation of this difference is the likelihood of greater fear of homosexual tendencies on the part of the male. They quote Sears's statement: "Sex anxiety is essentially heterosexual in females, but is essentially homosexual in males."[16] In dress, for example, females (both child and adult) can cross-dress (in pants) but males cannot. Likewise, dolls and tea sets for boys are viewed as much more aberrant toy choices than trucks and airplanes for girls. Fathers are often seen as especially sensitive to cross-sex preferences in their sons. A rather interesting incident occurred to me once that illustrates how widely held this view is. I was shopping for a shower gift for a pregnant friend. I ruled out pink myself, thinking of the social inappropriateness if a boy were born, and was concentrating on clothes in blue (acceptable for girls as well) and the neutral yellow. I was considering a blue outfit that had a few ruffles (but that could hardly have been called frilly or feminine) when the saleswoman interjected: "I wouldn't buy that if I were you. If the baby's a boy, the *father* won't want him to wear that." In thinking it over, I decided she was probably right (given current sex role conceptions) and made another choice. The interesting thing to note is the woman's understanding of the nature of cross-sex role sanctioning: it was the father she mentioned rather than the mother (who would have been mentioned in almost any other infant care context).

In modern-day America, we have one more complicating factor to consider in the area of differential parental treatment, and that is the role of infant and child care manuals. Every society will have a repository of collective wisdom about proper modes of child care, but modern American society is unique in its acceptance of the word of written authority in this area. The most prominent example that comes ot mind is, of course, that of Dr. Benjamin Spock's *Baby and Child Care,* which has sold 28 million copies since the first edition was published in 1946 (just in time for the postwar baby boom). In the first two editions, Spock described the tendency of girls to identify with their mothers by playing with dolls and other domestic toys, while boys were preparing for their adult roles by identifying with their fathers and playing with male sex-typed toys. He also disapproved of working mothers and always referred to the baby or child as "he." Spock has been roundly criticized for these views, which have been seen as supports for the traditional sex role system. In a third ediction, which appeared in 1976, Spock revised many of these views (including removal of "he" as the universal case), and in fact explicitly stated: "The main reason for this third edition . . . is to eliminate the sexist biases of the sort that help to create and perpetuate discrimination against girls and women."[17] As an aside, it might also be noted that in the third edition, Spock acknowledged for the first time the considerable help of his wife in the writing of all editions of the book.

In addition to the influence of written manuals, parents are also subject to the (often unsolicited) advice of relatives and friends, which may have relevance to sex roles. Grandparents may introject their own sex role views by choosing sex-typed gifts as well as by airing their own views directly. Parents may compare notes with other parents of same-age and same-sex children and come up with a repository of "facts" about what to expect in boys and girls.

Although there is considerable talk about "nonsexist" socialization, it would seem that most parents still have definite ideas about sex role deviance. Even among more reform-minded parents, the push toward nonsexist upbringing seems to be strongest in the case of girls, with boys receiving a more traditional sex-typing experience (perhaps because of the anxieties about homosexuality alluded to earlier). Given the reality of sex roles as a structuring element in society and in the person, it would be most difficult to raise a child without reference to sex roles at all. Although there can be considerable latitude in interpreting sex role constraints, it would seem that, given current social reality, a truly sex-blind socialization is an impossibility.

Identification

Most theories of sex role development save a central place for the concept of identification. It is a parsimonious idea for explaining the origin of the multitude of associated behaviors and attitudes that make up sex roles. It is dependent on the previous existence of an ongoing culture bifurcated by sex roles, which has created parental models for the child to emulate. Identification has been defined as "an acquired, cognitive response within a person. The content of this response is that some of the attributes, motives, characteristics, and affective states of a model are part of the person's psychological organization."[18] This definition could well be extended to include emotional and motivational responses as well, all acting to structure the individual's social psychological world. For the child in a nuclear family, of course, the chief models are the parents, but these models are reinforced by the existence of general cultural models for sex role behavior made accessible through the media, the schools, and other sources. Most theories, however, have put the primary emphasis on the parents, and we shall consider several theories of sex role identification that feature the parents in central roles.

Psychoanalytic Model

The Freudian view of sex role development has been very influential, both in its positive and negative impacts. It has influenced the thinking of many psychologists and inflamed the passions of many feminists. Let us first turn to a general consideration of the Freudian system, and then address more specifically the issue of sex role differentiation.

Freud took a basically biological approach to mental life. He sought to ground his concepts in the powerful instinctive life of the individual, a shadow of the biological imperative of the species to reproduce and survive. The id is seen as the basic personality system, the repository for instinctual energies and, ultimately, the source of all psychic energy. Out of the necessity of dealing with the world, the ego and superego systems are created, but important aspects of both of these processes as well are tied to the id and are inaccessible to consciousness. Sex role development grows out of this shadowy interplay between personality systems, but it is ultimately seen as a biological process:

> We are faced here by the great enigma of the biological fact of the duality of the sexes; it is an ultimate fact of our knowledge, it defies every attempt to trace it back to something else. Psychoanalysis has contributed nothing to clearing up this problem which clearly falls wholly within the province of biology. In mental health we only find reflections of this great antithesis; and their interpretation is made more difficult by the fact, long suspected, that no individual is limited to the modes of reaction of a single sex, but always finds some room for those of the opposite one. . . .[19]

Several general aspects of the Freudian view are salient for understanding the psychoanalytic perspective on sex role development. First of all, the centrality of sexuality as an organizing principle in mental life; second, the enduring impact of childhood emotions and perceptions on adult life, and third, the salience of the "politics" of the nuclear family for psychological functioning and development. Overriding all these factors is the pervasive one of psychic determinism, the fact that every aspect of human behavior is determined by psychological motives, most of them of unconscious origin. Thus, the rationale for much, if not most, of human behavior is thought to be inaccessible to consciousness, so that the choices made are motivated by psychic reasons hidden deep within the person's childhood experiences.

Strictly speaking, there is no such thing as a "sex role" for Freud. Sexuality is an integral part of psychic being and social functioning and cannot be seen separately from other parts of life. The choices made as to career, spouse, and life plan spring out of the unconscious maelstrom of childhood-based conflicts and motives, and relate only tangentially to the surrounding social reality. Since sexuality is so central, however, there are some commonalities of development within each sex, because of common experiences within the tight network of the nuclear family. The basic fact of the family structure that is most salient for psychic development (especially the development of the male) is the existence of strong sexual desire between child and parent. The desire for incest is demonstrated for Freud in the universal prevalence of the taboo against incest between mother and son, father and daughter, and brother and sister. It is the first two of this triad of possible incestuous relationships that most interests Freud, especially the first: the mother-son case, celebrated in the Oedipus myth. Freud argues that if the desire for incest did not exist, there would be no need to forbid it; thus the strong prohibition and strong sanctions accompanying it are testimony to the power of the urge.

The son, then, desires sexual union with the mother (a generalized sexual desire, not necessarily one exclusively for intercourse). This desire presumably arises out of the close caretaking bond that has developed between mother and son, as well as the beginnings of the heterosexual sex motive. The son's longing becomes most prominent at about ages four to six, when phallic sensitivity is the emerging erogenous zone. However, at the same time, the boy begins to experience guilt over these desires, as well as fear of the father's retribution for these sexual feelings. In the child's mind, this retribution takes the form of castration, creating what Freud calls a condition of castration anxiety. This anxiety motivates the giving up of the mother as the sexual object and the taking of the father as the identification figure, specifically with regard to the superego, the repository of moral values. The resolution of the Oedipus complex, as shown in the successful development of the superego, is seen as the single greatest developmental step in the life of the boy. The boy never really loses his love for his mother, however, and at sexual maturity, his choice of a wife is presumably guided by her similarity to the maternal model.

With reference to sex role identity for males, what Freud proposes is a possible explanation for the taking of the father as the identification figure, which would presumably motivate the acquisition of masculine values and traits, as well as the masculine-tinged superego of norms and prohibitions, specifically in regard to sexual repression. The boy, then, essentially identifies with the aggressor out of an emotionally driven fear of castration, which grows out of the original love link with the mother. Since the closeness of the mother-child link and the inevitable romantic triangle thus created seem to have some plausibility in their application to the nuclear family, Freud has at least provided a reasonable setting for the formation of male sex role identity. Whether the process takes place as Freud argued and can in fact have such long-range effects throughout the lifespan of the boy is more problematic.

In any case, then, male sex role development is primarily seen in terms of identification, a familiar enough concept from the psychological literature. Moreover, a very strong motivating force to support this identification process is also posited. In the case of female development, however, the process becomes more complex in the Freudian view. The girl's original love object is also the mother, but because of her sex, she does not pose a threat to the father. Besides which, since the girl lacks a penis, there is no punishment that could presumably be inflicted on her that would motivate identification. Two problems then exist in the Freudian view of female sex role identity: (1) the transfer of affection from the mother to the father; (2) the taking of the mother as the primary identification figure. Instead of appealing to an entirely new process, Freud seems to have preferred to "adjust" the male case to fit the female situation. He therefore argues that at the phallic stage, the girl first discovers the genital difference between the sexes and evaluates her position most unfavorably. That is, she feels "castrated," just as a boy might feel at the "loss" of his organ. Of course, in her case, it was never there to begin with, but Freud seems to think that interpersonal comparison can lead to the same results as intrapersonal comparison.

In Freud's view, the girl then covets the "missing" organ, resulting in penis envy. Because of her lamentable anatomical state, she develops a lifelong feeling of inferiority. As Freud puts it:

> After a woman has become aware of the wound to her narcissism, she develops, like a scar, a sense of inferiority. When she has passed beyond her first attempt at explaining her lack of a penis as being a punishment personal to herself and has realized that that sexual character is a universal one, she beings to share the contempt felt by men for a sex which is the lesser in so important a respect, and, at least in holding that opinion, insists on being like a man.[20]

It is, of course, statements like this that have angered feminist critics of Freud. Before considering their views, let us complete our discussion of his position. The girl, then, experiences the castration complex, the feeling of already being castrated, and this dominates her reactions to men, women, and herself; all are evaluated in terms of possession of the valued organ. When she discovers that the mother also lacks a penis, she blames the mother for her own state and therefore rejects her as a love object, turning to the father instead. This establishes the basis for later heterosexual attraction: the desire to have the penis, and ultimately to give birth to a son as the symbolic fulfillment of that wish.

The theory has so far attempted to explain the choice of the male love object, but we have yet to understand the girl's choice of the mother as her identification figure, a crucial step in the development of sex role identity. This choice, according to the Freudian scheme, is made when the girl realizes the futility of having her father and fears the consequent loss of the affection of the mother, on whom she is still very much dependent. Consequently, she takes her mother as her identification figure, but this is never as complete a process as it is in males, nor is the superego developed satisfactorily either. The reason for the incompleteness of the process is that "the motive for the demolition of the Oedipus complex is lacking," that is, since the girl cannot fear castration, she feels as if she has already been castrated: there is nothing left for her to fear. Consequently, Freud argues, there are many cases of incomplete feminine development, mostly because of the sheer difficulty of completing the process as conceived here. Freud comments on this:

> A man of about thirty strikes us as a somewhat unformed individual, whom we expect to make powerful use of the possibilities for development opened up to him by analysis. A woman of the same age, however, often frightens us by her psychical rigidity and unchangeability. Her libido has taken up final positions and seems incapable of exchanging them for others. There are no paths open to further development; it is as though the whole process had already run its course and remains thenceforward insusceptible to influence—as though indeed, the difficult development to femininity had exhausted the possibilities of the person concerned.[21]

By reading such accounts of feminine development we can understand the fury these views arouse among feminist critics. However, if we delve a bit deeper into Freud, we can see that his view of sex role development was considerably more complex than some give him credit for. A constant theme in his work, for example, was the bisexuality of all men and women, with the terms "masculine" and "feminine" being in some sense oversimplifications of a much more complex reality. He wrote:

> . . . in human beings pure masculinity or femininity is not to be found either in the psychological or biological sense. Every individual on the contrary displays a mixture of the character traits belonging to his own and to the opposite sex; and he shows a combination of activity and passivity whether or not these last character traits tally with his biological ones.[22]

Freud maintained that psychosexual development is essentially identical for the two sexes up to the phallic stage (ages four to six), so that an initial period of bisexuality underlies all subsequent development. He also recognized the contribution of social factors to the development of sex roles, an area he is often accused of neglecting. He wrote:

> But we must beware in this of underestimating the influence of social customs, which similarly force women into passive situations . . . The suppression of women's aggressiveness which is prescribed for them constitutionally *and imposed on them socially* favors the development of powerful masochistic impulses . . .[23] (emphasis added).

Nevertheless, Freud did enunciate a definite view of female development, indicating that the girl's anatomy and inevitable reaction to her lack of a penis lay the groundwork for a very different psychic character than that of men. Passivity is a key element in female character, Freud believed, as are unresolved feelings of inferiority because of the female anatomical condition. Helene Deutsch later elaborated on Freud's views, emphasizing the traits of passivity, masochism, and narcissism as the components of the "feminine core." Passivity was woman's fate because of her constitutionally less active instincts, her "inadequate" clitoris, and the inhibitions taught her by the family. Aggression was thus turned inward in women, according to Deutsch, and led to masochism, the desire to subject oneself to the will of others. Narcissism arises as a compensation for felt sexual inferiority: the physical self is overvalued, leading to vanity in women. Deutsch focuses on the normality of passivity and its centrality for femininity. Sexuality was, in her view, to be expressed primarily through motherhood and identification with a male sheltering figure. Clitoral sexuality was to give way to vaginal sensitivity but such a transition was in fact not usually attained.

Surprisingly enough, Deutsch rejected the idea of penis envy as a primary motivating force for female character traits. Instead she favored an explanation of constitutionally more passive instincts, as well as the childhood inadequacy of the clitoris as a masturbation outlet. The girl turns to the father as an active agent who presents a model for ego mastery, but her development is always circumscribed by her constitutional limits, and she is happiest when living within them. Deutsch spent a good deal of time cautioning against the dangers of cold intellectuality, the trap of the "masculinity complex":

> Woman's intellectuality is to a large extent paid for by the loss of valuable feminine qualities . . . it feeds on the sap of the affective life and results in impoverishment of this life.[24]

To complete the picture of biological determinism, Deutsch did not give much credence to social factors as determinants of female character. Instead, she maintained:

> While we recognize the importance of social factors, we assume that certain feminine psychic manifestations are constant and are subject to cultural influences only to the extent that now one and now another of their aspects is intensified.[25]

Deutsch's work built upon Freud's notions of femininity, but emphasized biological factors to a greater degree. Her work and Freud's were later popularized by Lundberg and Farnham in *Modern Woman: The Lost Sex,*[26] which subsequently became a prime target of feminist critics. In a recent interview (the original work was published in the 1940s), Helene Deutsch came out in favor of career interests for women and a more active model of femininity. She remarked:

> Here there is a misunderstanding. Passive and receptive refers to the instinctual life—the biological drives, the nature of sexual activity, should be passive but the ego can be active in both men and women. The ego can be whatever it will.[27]

Although revision and explanation of psychoanalytic theories of female development have softened the impact of some of the concepts, they still provide plenty of ammunition for critics. Three major points have aroused special criticism: (1) penis envy, (2) vaginal sexuality, and (3) the constitutional nature of "feminine" traits as opposed to cultural influences.

(1) Penis Envy. Karen Horney[28] has maintained that the concept of penis envy is a male-oriented one. That is, although it is presumably an emotional judgment made by little girls, penis envy actually corresponds to the male evaluation of his organ and his view of women. Horney thinks it is unlikely that girls put the same evaluation on the penis as males do. Instead she argues that there is a male envy of the female capacity for motherhood. Male depreciation of the female role may be seen as a reaction to this felt inferiority. Adult female recall of penis envy during psychoanalysis may simply reflect secondary reactions to difficulties in female development, including the obvious one of "the actual disadvantage under which women labor in social life."[29]

Clara Thompson also has criticized the concept of penis envy in women, arguing that cultural factors can account for clinical findings. She elaborates:

> Therefore, two situations in the culture are of importance in this discussion: the general tendency to be competitive which stimulates envy; and the tendency to place an inferior evaluation on women. No one altogether misses some indoctrination on these two trends. If the competitive attitude is greatly developed by personal life experiences, the hatred of being a woman is correspondingly increased. The reverse is also true; that is, if there has been emphasis on the disadvantages of being a woman, a competitive attitude toward men tends to develop. Out of either situation may appear character developments that fit into the clinical picture of penis envy, and it is not necessary to postulate that in each case an early childhood traumatic comparison of genital organs took place.[30]

(2) Vaginal Orgasm. Due to the centrality of sex for Freud, the issue of vaginal vs. clitoral sensitivity also has relevance for sex role socialization. Especially as elaborated by Helene Deutsch, the passive aims of vaginal sexuality become an organizing principle of female character as well as a desired end state for achieving sexual satisfaction. Many feminist critics[31] have taken aim at this view, citing research by Masters and Johnson[32] that seems to demonstrate that female sexual sensitivity resides solely in the clitoris (with little vaginal sensitivity), since the clitoris is the site of nerve endings and corresponds to the male penis. In female orgasm, the clitoris takes an active role, although some vaginal contractions usually occur as well. No physiological or neurological distinction can be drawn, then, between a clitoral and vaginal orgasm. In fact, Masters and Johnson's research seems to emphasize the similarities

between male and female sexual experience, focusing on the active components of each. The hypothesized active-passive distinction between male and female sexuality is a basic requirement for the psychoanalytic conception of sex roles. One could argue, of course, that the psychical consequences of male and female sexual experience are different even though this difference does not seem to register on sophisticated scientific monitoring equipment.

Freud's view of the difference between male and female sexuality also corresponds to the well-known double standard, in which the drive toward and enjoyment of sex are mainly seen as male prerogatives, with female sexuality subordinated to the emotional level of the interpersonal relationship. In the Victorian era, with its repression of sexuality and idealized image of woman, this view was the accepted one, and it no doubt influenced Freud. Some recent research by the historian Carl Degler[33] has cast doubt on this notion of Victorian sexuality, arguing that nineteenth century physicians and other commentators recognized the existence of female sexuality, the female orgasm and the essential similarity between male and female sexual experience.

(3) Biological vs. Cultural Origin of Female Traits. Finally, criticism of the psychoanalytic perspective has noted the probable cultural genesis of female traits heretofore ascribed to biological factors. Horney, Thompson, and others support this view. The Freudian rejoinder would of course be to question the origins of these cultural biases and to trace them back to biological factors. The argument would also be made that only biology could account for the uniformity across time and space of sex role characteristics, a position that is challenged by many social scientists.

Juliet Mitchell,[34] a sympathetic feminist commentator on Freud, has argued that the feminist attack on Freud is based on a fundamental denial of the existence of an unconscious mental life. She then goes on to support this view by reference to dreams, fantasies, and neurotic symptoms, the visible tip of the iceberg of mental life. According to Mitchell, social reality only gains its power through its incorporation into the unconscious of the individual. The power structure of patriarchal culture may be the ultimate ground for such images, but the images themselves take on an independent being and power in the psychic life of the individual that cannot be discounted.

What can we conclude, then, of the validity of Freud's view of sex role development? On the positive side, he provides a complex motivational structure on which to base the identification process. Even within the bounds of his system, however, the plausibility of the female plan of development is quite strained. On the male side, evidence for the Oedipus complex and castration anxiety is drawn mainly from psychoanalytic sources, as the nature of the theory does not readily lend itself to experimental verification. Acceptance of the Freudian view of sex roles requires belief in the primacy of the unconscious, the central role of sexuality, and the enduring effect of childhood on adult personality, all bases of the Freudian system.

Freud has captured something about the quality of psychic life that has eluded other commentators, and it is this essence of truth that makes one loath to dismiss his entire system on the basis of conflicting evidence on cultural inputs and the nature of sexuality. Some theoretical conceptions, like penis envy, also give us pause, and the entire substance of the theory seems alien to more rational views of psychological functioning. We do not know how to probe the unconscious (if we agree it exists and is important) without the guiding light of theoretical conceptions, though these very conceptions may blind us to the extent and variety of what we might find there and force our observations into prematurely rigid categories. In a later chapter on symbolism, we will again meet the unconscious in terms of its symbolic representations, but for the time being we shall leave it and move on to alternate views of identification, ones that do not emphasize the unconscious roots of behavior.

The two views of identification we consider next are the social learning model and the cognitive-developmental. Taken with the first, the psychodynamic Freudian model, these perspectives on identification mirror three of the most influential theoretical conceptions in contemporary psychological thought. The social learning model derives ultimately from the behavioristic point of view, which emphasizes the importance of behavioral outcomes, particularly reinforcing ones, for the imprinting of behavioral patterns. Personality traits are seen as packages of behavioral tendencies with little emphasis on introspective psychological experience. The cognitive view, on the other hand, insists on the primacy of cognitive representation. Piaget's work is perhaps best known in this field and is the forerunner of Kohlberg's thinking. The cognitive view harks back to an older, introspectionist theme in psychology. The dynamic psychoanalytic perspective is, of course, also one that emphasizes mental life, but largely of the unconscious sort that is not directly addressed by cognitive theorists. We shall see, though, that for Kohlberg the cognitive schema of sex roles are not affectively neutral, so that the two realms do meet in this theory.

Social Learning Model

The social learning model[35] is not exclusively behavioristic, since an auxiliary mechanism to reward and punishment has been offered: that of imitation and modeling. However, the choice of model is premised on the reinforcing qualities of the model. Social learning theorists point out the similarity between the psychodynamic conception of identification and the concept of imitation. Both involve observational learning, in that behaviors are acquired through observation of a model engaging in them. To demonstrate instances of observational learning, researchers set up studies in which models (usually adults) are made available to the child, and a tally is then made of the behaviors imitated by the child. Such studies have shown that nurturant and powerful models tend to be the most effective in inducing imitative responses. Parents, of course, have both of these qualities and are accessible to the child over a long period of time.

The concept of identification as used in psychodynamic models seems to have a holistic quality that is missing in the social learning view. Although cognitive and affective aspects of modeling are mentioned in the theory, the core concept seems to be the learning of discrete responses. Shaping of behavior also takes place, in that children are directly rewarded or punished for behaviors that are appropriate or inappropriate to their roles, and linguistic labels, like "big boy" and "sissy" can acquire secondary reinforcement qualities that aid in giving the child a sense of what is expected. The child then generalizes from his or her limited experience to broader domains of behavior, generating guidelines for living that we might term sex roles. The child learns that behaviors are differentially appropriate to the two sexes through direct reward and punishment and by observation of live models (e.g., parents) and symbolic ones (e.g., those in the media). The labels "boy" and "girl" are used often in socialization and are very important in teaching sex roles, since they indicate to the child which behaviors if enacted will most likely result in reward or punishment (the guiding principles behind any behavioristic notion of response acquisition).

Aggression has received much scrutiny from social learning researchers. It is of course of prime importance in sex role socialization. Bandura[36] reports on a study of children observing an aggressive model, in which imitative aggression was higher for boys. When a reward was then offered to all children for performing the aggressive response, the sex difference was wiped out, with girls performing as well as boys. Bandura hypothesizes that the original sex difference was caused by the differential reinforcement histories of the two sexes, but was eliminated by the offer of the reward for performing the response. Both sexes may be competent in their knowledge of aggressive responses, but may differ in their performance of them because of what their past experience and knowledge of sex role standards leads them to anticipate.

Jerome Kagan[37] has offered a model of sex role development that seems to combine aspects of the psychodynamic, cognitive, and social learning formulations. For Kagan the central concept is that of the sex role standard, defined as "a learned association between selected attributes, behaviors, and attitudes, on the one hand, and the concepts male and female, on the other."[38] The child learns to discriminate between the categories male and female, aided by the culture's considerable concern and investment in this division as transmitted by parents and other socialization agents. Once this discrimination is made, it lays the groundwork for the acquisition of sex-typed responses. The child acquires many sex-typed skills through social learning, as indicated above, but also identifies with the same-sexed parent because of the "desire to command the attractive goals possessed by the model."[39] The child thinks that if he or she were to act like the model, then he or she would also acquire the positive qualities of the model (power, positive regard of others). The sex role standard is inculcated by the parents and by the culture and acts as a cognitive-emotional guide to appropriate behaviors. The child acts from what Kagan calls a fundamental human motive: "the desire to make one's behavior conform to a previously acquired standard."[40] Acting in accordance with that standard is self-reinforcing to the child and tends to inculcate sex-typed behaviors and feelings more deeply. Thus Kagan adds a cognitive and affective dimension to the social learning paradigm, altering it considerably in the process. Reinforcement, however, is still seen as a prime determinant of behavior, albeit the concept has been extended to include self-reinforcement as a consequence of actions in accord with an internal standard. That standard gains its power through the reinforcing consequences of acting in accordance with it. The idea of a sex role standard does indicate the presence of some sort of abstract schemata that acts as a cognitive guide for behavior, a concept that sometimes seems implicit in social learning theory but that is not made explicit. Without the notion of some sort of unifying guide for behavior, social learning theory suffers from the fragmentary nature of its hypothesized responses, which never seem to cohere into the more definitive mold of sex roles. Walter Emmerich has argued that "adult sex roles in modern society have neither the coherence nor permanence required to serve as specific targets of early socialization."[41] Nevertheless, most theorists act as if sex roles have a reality that extends down to influencing the course of childhood socialization both in terms of plans for present behavior and anticipations of future adult roles.

Models of sex role acquisition that depend on reinforcement and imitation do not suffer from lack of empirical support. Laboratory experiments providing evidence for this position are relatively easy to set up, as they involve measures of discrete behavioral responses. In addition, the concepts strike a chord in everyday experience, since we are all familiar with instances of children's imitation of adult models and responsiveness to reinforcement. However, strict belief in the social learning position requires something more: adherence to an environmental view of behavioral causation and exclusion of more sophisticated notions of cognitive schemata and affective motivation. As mentioned previously, there are rudimentary aspects of both cognitive and affective factors incorporated in the theory, but they are definitely secondary to the fundamental principles of learning. The child is viewed as a tabula rasa, ready to be imprinted by the contingencies of reinforcement, and presumably subject to change if such environmental inputs were to be altered. Traits as organizing foci for personality are all but absent, replaced by a network of reinforcement associations drawn from direct experience, observational learning, and vicarious reinforcement. The controversy about social learning theory, then, does not so much concern denial of its role in sex role acquisition, but rather its image of the functioning personality structure of the individual in interaction with the environment. Also in dispute is the idea that reinforcement associations are sufficient to account for the strength and tenacity of sex role identity and for all aspects of it, including not only externally observable behaviors but feelings and motives as well. The adequacy of such a model as the sole explanation is in doubt, but few would dispute the great importance of learning principles in sex role acquisition.

Cognitive-Developmental Model

Lawrence Kohlberg[42] is the author of a theory of sex role development that draws its power from the image of a motivating cognitive judgment made early in the child's life that leads to identifications and the performance of sex-appropriate behaviors. The child makes an unalterable cognitive categorization of himself or herself as boy or girl, and this judgment then organizes the subsequent development of behaviors. According to Kohlberg, "because gender is the only fixed general category into which the child can sort itself and others, it takes on tremendous importance in organizing the child's social perceptions and actions."[43] The child begins to make such a categorization at about age four, but the process is not really complete until the age of six or seven, when the link to genital differences is established.

According to this theory, the child's understanding of sex roles is limited by his or her cognitive ability to understand the world. Piaget has developed a theory of universal stages in cognitive development in terms of which the child's understanding of physical objects and persons is processed. Kohlberg says that children comprehend sex roles in physical terms, that is, in terms of body size and shape differences, so that their understanding of sex roles is linked to their ability to understand physical objects in general. Piagetian research has indicated that up to a certain age, children are incapable of ascribing permanency to physical objects, and this response also includes sex roles. Therefore, young children think that if a boy changed his appearance he could become a girl, and that a girl could do likewise. The full cognitive understanding of the permanence of gender is a requirement for sex role development, and does not fully develop until the age of six or seven.

Once this final cognitive judgment is made, it acts as an organizing focus for future behaviors. The child's self-categorization leads him or her to value same-sexed behaviors and objects positively. The girl, then, can be seen as saying "I am a girl; therefore I want to do girl things." Doing "girl" things becomes rewarding in itself, as it accords with the cognitive judgment of self. The child's positive evaluation of the self thus extends into the outer world to affectively color behavior and objects in accord with it. The child will then seek out situations and models that are in accord with this cognitive categorization. Chief among the models found in the child's everyday environment are the parents. Thus the child becomes attached to the same-sexed parent and imitates him or her because of the association with the child's own sex. For social learning theorists, attachment precedes modeling; for cognitive theorists, modeling precedes attachment and is controlled by the child's cognitive judgment of self.

Evidence for Kohlberg's theory has been fragmentary, although its major points are in accord with findings by other investigators. For example, research by Money and the Hampsons[44] has indicated that when sex reassignment is attempted after the ages of two to four years, it usually cannot be successfully engineered since the child has already developed the rudiments of a sex role identity. Before that critical period, successful sex reassignment is usually possible. Kohlberg hypothesizes that cognitive constancy of the gender concept takes shape at about that period, so that the child is then resistant to change in it. Before that time the child's sense of gender is very rudimentary, since the cognitive organization of children of that age does not permit full grasp of the invariance of physical objects or concepts.

Kohlberg observes that by age six or seven, the child usually exhibits a strong gender identity, making sex-typed choices and usually preferring same-sex peers. At this age the child is cognitively able to perform concrete operations in terms of fixed logical classes, an ability that is reflected in the constancy and fullness of gender role. Also at this age, there develops in most children a marked tendency to evaluate their own sex and sex-typed activities positively. This trend is especially true for boys, since at this age judgments of value are often confounded with judgments of superior size and strength, which are usually male attributes. At later ages there is a transition to a more abstract notion of sex roles, paralleling the general cognitive trend in the child. Sex roles are then linked to the social system in terms of work and family roles, and at a later time may relate to the person's own conceptions of what he or she wishes to be.

Little specific research has been done to validate the cognitive-developmental conception. Especially critical in the system is the sequence of (1) gender role concept; (2) positive evaluation of self, and by extension, sex-typed persons and activities; and (3) identification with the same-sexed parent. Since the early understanding of the child is so bound up with issues of physical size and strength, it makes sense to suppose that boys grow to think "bigger and stronger is better" and generalize this positive affect to sex-typed persons and activities. What of girls? Unfortunately Kohlberg gives male examples and has only reported interviews of males to support his theory, thereby following the unfortunate precedent of many researchers of considering the male to be the standard case and the one most worth looking at. It is likely, though, that girls also learn to evaluate their sex positively, although a discordant note is present in their knowledge of boys' negative evaluation of them (i.e., the male idea that men are bigger and stronger and smarter obviously implies a negative image of females, one that is often enunciated rather readily by six-year-old male chauvinists). But, unlike Kohlberg's idea of the physical basis for sex role concepts and positive regard, which seems to fit the male case but not the female, it is likely that girls gain a positive image of their sex by association with their mothers in their nurturing child-rearing role. Exposure to females is very high for children in any case and may make a special impression on girls because of their growing sex role concept, which facilitates a feeling of kinship with their mothers. Girls, then, may observe the physical size and strength differences as well as boys, but these are probably not as salient to them in their development of positive regard for sex-typed activities. A more potent influence on their development is probably the positive aura surrounding nurturing experiences with their mother and other female figures. Exposure to dolls and domestic playthings (such as tea sets and doll houses) is a likely facilitator of the process as well. The fact that tomboyish behavior is not an uncommon aspect of early female development probably indicates that girls as well as boys observe the physically based size-strength distinction as a basis for the positive evaluation of the male world seen in its terms.

The cognitive-developmental view, then, is inviting because it provides a grounding for a cohesive concept of sex role that acts to organize the child's perceptions of and attachments to the world, and in particular to parental figures and activities. Such a theory does not exclude the input of social-learning-based reinforcements and imitations, but it subordinates them to an immutable cognitive categorization that is central to the child's functioning in the world. Kagan's view, presented earlier, is somewhat similar in its description of an internally based organizing principle, but it is not specifically linked to cognitive stages and processes as is Kohlberg's. The cognitive-developmental perspective has not fully justified its image of sex role concept development in the female case. It favors a "unisex" process with no sex distinctions in it (unlike the Freudian system), but it runs into trouble in its attempt to base female sex roles and positive feelings in a male-based size-and-strength evaluation system.

Another system of conceiving sex role identity has taken the tack of specifying different identification paths for boys and girls based on the differing nature of what the mother and father provide to the same-sex child in the way of behavior that can be imitated. This conception, developed by David Lynn,[45] maintains that the father provides a rather abstract identification goal for the boy since he is away from the home much of the time and since for the most part the masculine sex-typed activities are not directly observable by the child. On the other hand, the mother-daughter relationship is one of close observational learning; therefore the girl learns to identify with a particular person—her mother—rather than with an abstract role. The boy first identifies with the mother as well, but because of his feeling of similarity to the father, he is gradually weaned from this identification and turns to the more abstractly defined role of the father. The boy retains some vestiges of the maternal identification, notably those that do not conflict with the masculine role. The girl never has to give up her original, personal identification with the mother. The consequences of these different kinds of identification show up in the differing cognitive styles of the two sexes. The boy must learn early to abstract material in order to learn his masculine role, whereas the girl can model herself directly on a present model within the context of a warm interpersonal relationship. According to Lynn, this early training in abstraction may be at least partly responsible for male superiority at abstract, conceptual tasks. This is a rather speculative hypothesis since the performance differences can also be explained (perhaps more directly) by differential reinforcement and experience with conceptual areas, as well as by the definite male sex-typing of such tasks themselves. In any case, we do not have to follow Lynn's thinking to its furthest extreme to note that he has pointed out something of value: the decided difference between a boy's identification with his father and a girl's identification with her mother.

A Final Note on the Role of Parental Figures

The three theories we have discussed, the psychoanalytic, the social learning, and the cognitive-developmental, all give central roles to parents as transmitters of societally based sex roles. Because of their continued presence and their great emotional importance to the child, parents play a pivotal role in the establishment of sex role identity. Such socialization, however, is only meaningfully understood in the context of a societally based view of sex role concepts. Parents, acting as models for their children, reflect the effects of their own sex role socialization and their continued exposure to a world that makes important social distinctions on the basis of sex. In a climate of changing sex roles, the parental attitude is probably one of compromise between the deeply ingrained patterns of the parents' own childhoods and sex role identities and the pressures of the contemporary world. Parents provide a conservative link to the past in that many of their behaviors are shadows of a distant socialization experience of their own.

Most theories of sex role identification give major emphasis to the same-sex parental model. Note might be made here of the important role of complementary identification. The child also learns the other sex role by observation of cross-sex parental behaviors and by participating in cross-sex parent-child relationships. We have already seen evidence for a more permissive flavor to cross-sex than same-sex parent-child relationships, and, of course, the psychoanalytically based perspective would strongly affirm the nature and importance of this link. A girl, for example, could well be more encouraged in feminine-stereotyped behavior by her father's attitude than by her mother's. Outside of the family as well, cross-sex relationships (both adult-child and child-child) may be critical in patterning sex role behaviors. In fact, much of what we understand by the terms "masculine" and "feminine" we define in terms of cross-sex behaviors.

Other persons in the home may also have a strong effect on sex role socialization. Siblings are the most obvious example. Older siblings, in particular, may act as surrogate parental models for sex role behaviors. Evidence has shown that boys with older sisters have somewhat more "feminine" interests (defined in terms of the stereotyped masculinity-femininity notions to be discussed later in this chapter) than do boys with older brothers. In the modern case, the typical nuclear family will only include parents and children, but historically this is the exceptional case. Most societies (including earlier American cases) are based on the extended family, with grandparents and sometimes other adult relatives commonly present. In such cases, the child is provided with a series of adult models and identification figures, although it is still likely that the child's parents retain the primary role in the socialization of sex role identity.

The three major identification theories emphasize different mechanisms: affective bonds, modeling, and cognitive categorizations. It is likely that all three mechanisms play decisive roles in the process of sex role socialization. The child must develop a cognitive grasp of what it is to be a girl or a boy; this can hardly be avoided given the preponderance of overt sex role messages that bombard the child. Such messages gain their salience in the close emotional context of the family in which ample opportunity for identification and modeling occurs. The familial sex role context is only the threshold experience for the child's sex role identity, however; she or he has yet to be immersed in the school environment, peer group culture, and the symbolic world, all of which complement and usually reinforce the initial familial experience. The lasting impact of the family on sex role socialization is difficult to dispute, however. All of us carry in the structure of our personalities and visions of the world this imprint of our earliest and most lasting message about our roles as men and women.

SOCIALIZING EFFECTS OF THE SCHOOL ENVIRONMENT

The influence of the school environment on sex role socialization can be considerable.[46] The teacher can of course directly transmit sex role expectancies by specifically indicating his or her feelings about appropriate and inappropriate behaviors. For example, in presenting material on occupations, especially in the context of guidance counseling, very definite sex-typed expectations can be transmitted. The teacher could, for example, discourage girls from seeking professional careers either directly by disapproval or emphasis on marriage and child care as an alternative, or indirectly, by encouraging only boys in such career choices. In the early grades, when presenting material on men and women, it is only natural to expect that the teacher will present such material through the screen of his or her own view of the world. When there is choice in the material to be taught, a teacher who is open about sex roles could stress ordinarily hidden facts about the role of women in history, or take the usual and conventional path and teach a male-oriented view of the world. After sitting through years and years of lessons that only rarely mention women (or only mention them in stereotyped contexts), it is no wonder than many students take with them from school a sex-stereotyped view of life. Of course, the teacher is ordinarily constrained in what he or she presents by the actual materials available, which are often highly sex-typed.

In the assignment of tasks in the schoolroom, e.g., moving furniture, and dusting, opportunities for sex-typing abound. The teacher may explicitly emphasize sex role distinctions by remarking on the qualities of girls in general or boys in general, as well as by treating boys and girls in different ways. Teachers sometimes specifically punish out-of-sex-role behavior and support in-role behaviors. This is not always done in overt and obvious ways. For example, Barbara Harrison presents the example of a kindergarten teacher supervising carpentry in a group of boys and girls. "A girl shows her teacher (male) her handiwork. 'These nails aren't hammered in far enough,' he says, 'I'll do it for you.' A boy shows the teacher his handiwork. 'These nails aren't hammered in far enough,' the teacher says; 'Take the hammer and pound them in all the way.'"[47] Instances like these can be multiplied many times in the life of the child and transmit a unified message about sex role standards. For example, higher levels of achievement in math and science may be expected from males than from females.

The ways teachers behave can be very subtle indeed, as was demonstrated by Louise Cherry's[48] analysis of teacher-child interactions, in which more attention-getting and controlling features were embodied in the teacher's speech to boys than to girls. Several studies have shown that boys do present more of a disciplinary problem in the classroom. Patricia Sexton has argued in her book *The Feminized Male* that the norms of the classroom (quiet, conformity) accord more closely with feminine than with masculine values, accounting for the school difficulties many boys have, especially in the early years. A study by Jerome Kagan has also demonstrated that children more often attribute female labels to common classroom objects than male labels, supporting Sexton's ideas.

The schoolroom itself is often organized in a stereotyped way, with a girls' doll corner and a boys' truck and block section. The curriculum itself may include teaching of stereotyped skills, such as sewing and carpentry, and restrict access to them on the basis of sex.

As in the area of teachers' expectations, the effect of content is often quite subtle and all-pervasive. Analysis of children's books indicates a heavy preponderance of male characters (often animals or fantasy creations are referred to as "he"). The content of such books is often a microcosm of sex role stereotypes. Dick and Jane and other readers often teach sex roles along with reading; female characters are shown in sex-typed activities and adult females are usually given only a family role, not an occupational one.

As has been mentioned previously, textbooks often ignore the contributions of women, so that even a teacher who wants to present a fair picture is faced with a dearth of materials to draw from. Some efforts are now being made in revising texts to include female achievements as well as those of minority groups. Further discussion of this problem will be found in the section on symbolic agents.

The school is a major locale for the operation of peer group influence, since many friendships and peer activities derive from the school setting. The world of the student peer group is not one that captures the attention of many educators, but it may be more far-reaching in its influence and success in enforcing conformity than is the classroom.

PEER GROUP EFFECTS

Most socialization research has concentrated on the parent-child dyad, yet the child spends a great deal of time with peers and is heavily influenced by them. Both at school and at play, the child is faced with a preexisting peer culture with its own values and activities, which may be heavily sex-typed in nature. Beginning at early school-age and continuing well into adolescence, the child is usually faced with a same-sex peer group to which he or she must become attached if friendship needs are to be met. Cross-sex friendships during these years are extremely uncommon and are often prevented by the peer group, especially in the case of males, who often specifically disdain female companionship. Regardless of the child's own attitudes, then, or those of parents, he or she is generally absorbed into association with same-sex peers and with a peer culture that tends to accentuate stereotyped features of sex roles.

For example, in the case of girls, peer activities in childhood are likely to include group doll play, playing house, or jumping rope. In contrast, male activities usually revolve around vigorous physical activities, war games, etc. Often organized groups (e.g., Little Leagues, which though nominally open to girls in some places remain a male-typed domain) accentuate the values of the group by institutionalizing them. The child is faced with an ongoing peer culture and must relate to it in some way or risk the fate of the social outcast. Such groups can be extremely coercive in promulgating acceptance to their norms and in punishing deviance by the threat of withdrawal of friendship. The peer groups need not be highly organized, e.g., into gangs, for these pressures to work, since the child simply has no alternative in seeking companionship outside of the home. The child, though, usually goes along willingly with the values of the group since they provide guidelines for behavior, and the values of the group are generally those upheld by the sex-stereotyped culture: i.e., aggressive, physical endeavor for boys; quiet, domestic activities for girls.

In recalling peer pressure in my own childhood, a few images come to mind. In the mid-fifties, a fad for so-called "Ginny dolls" occurred, involving the purchase of said doll and untold numbers of costumes for it (bridal, skating, dating, etc.), all neatly hung in Ginny's very own closet and dressing room. Although these dolls were actively promoted by television advertising, the real push to buy them came from the female peer group which for awhile revolved around the acquisition of as many fashion outfits as possible for the beloved Ginny. Although I also owned many non-sex-typed toys, e.g., electric trains, the sense of belonging and participation that owning this particular doll brought made it special. In fact, I can still recall the look of the doll in its brand new skating outfit, with its smart white ice skates. It is from such instances as these that a child's understanding of the world and future life patterns takes shape.

Male peer pressures in childhood often center around excellence in sports. The boy who is not athletically able is subject to social censure and often develops a poor image of himself as well. Both participation in sports and interest in televised sports are usually prominent features of the peer culture. Females are very definitely excluded from this realm, creating a "male culture" of sport that often continues, with less force, throughout adulthood. Girls also participate in sports, of course, but their participation is not centrally related to the female peer culture and nonparticipation is not usually censured.

The impact of the peer culture often extends into symbolic realms as well. Books to be read and heroes to be worshiped are often geared to peer group values. For example, as a child I recall the pressure to read books like *Sue Barton, Student Nurse* and *Little Women,* simply because everyone else was reading them. Such books reinforced the values of feminine domesticity and passivity. The books, of course, were not produced by the peer group, but it was peer pressure that led to the acceptance of such books and their values. The peer group, therefore, acts as a primary influence group for the processing and dispersal of sterotyped materials produced in the adult culture. Interest in sex-stereotyped hero figures is often mediated by the peer culture. Baseball stars, for example, gain their psychological reality for boys through their idealization in the peer culture.

Throughout childhood the peer group is an important mediator (and, often, exaggerator) of the sex-stereotyped values of the culture. For the most part, the child at any time is probably not aware of the power of peers since most children seek peer approval and act in accordance with peer norms. And in many cases the norms are not completely inflexible and can adapt to individual cases. In general, then, the child *wants* to behave in ways approved by his peers since that is the path of least resistance and greatest reward. Since peer culture generally only acts to magnify general cultural sex role norms, the child is likelty to be acting just about the way everyone expects if he or she goes along with the peer group.

It is not until adolescence that the hold of the peer group often becomes coercive and the relative fluidity of childhood peer relationships breaks down. Much research has been devoted to this period and its impact on the person and the social structure. A study by Bowerman and Kinch[49] indicated that in the early teens, adolescents turn away from their parents as primary models and guides for behavior and toward the peer group. The pressure for conformity to peer values can often be severe, especially for girls. A study by Coleman[50] of what he called "the adolescent society" found that physical attractiveness, popularity, clothes, and dating success were valued for girls, while scholastic success was devalued. Although these values are doubtless undergoing some change, remnants of them still remain. For boys, athletic and sexual success were found by Coleman to be most important, with scholastic achievement a possible secondary goal. Regional and social class differences could doubtless be noted in this value system.

Adolescence brings a narrowing of goals and a focusing on future life plans. Unfortunately, this narrowing is especially acute for young women. The emphasis on immediate social popularity noted in so many studies often takes precedence over academic goals and career planning. Although doubts about sexual attractiveness and competence plague both sexes, at adolescence such issues seem to obscure other goals for women, whereas they generally do not do so for most men. Douvan[51] studied 3,000 adolescents and found that the critical question for boys was: "What is my work?" while for girls it was "Who is my husband?" For some women who do attend college, peer group pressures for popularity with men often continue, and more attention is often focused on the weekend date than the lifetime career plan. Pressure for early marriage for women may be lessening now, but the primacy of relating to men instead of to work continues. The female peer group promotes no viable image of identity for women outside of definition through male relationships. Alternative views, promulgating individual assertion and achievement, are provided by some sources, such as women's movement reference groups for young women and magazines such as *Ms.*, but such values touch the lives of relatively few young women compared with the peer culture.

For young men, the seriousness of commitment to work soon overtakes the old peer values of athletic and sexual success, although the latter are never really forsaken completely. Men seem to retain much more of a sense of control and vision in terms of what their future life will be like than do women. The values of the adolescent and youth female peer culture dictate a timeline of life defined almost totally in terms of marriage as a goal. The timeline remains undelineated after that; it seems to lack content and reality after the critical step into the family system is taken. For men, on the other hand, the values of adolescent and youth peer culture do not take marriage as a stopping point but are oriented to the male role of supporting the family economically, with room for individual achievement and creativity in the more privileged economic classes.

Throughout life, same-sex peer groups tend to exert some influence, although this influence probably decreases after adolescence. Many occupations, both at the working class and middle class level, are heavily dominated by men and provide a peer reference group for maintaining sex-typed behaviors.[52] Women who do not work, especially those with children, are probably more isolated from peer support groups, although some are present in the form of organized women's clubs and service organizations (e.g., the PTA) as well as informal gatherings (such as playground mothers' groups, as well as women's consciousness groups at the other extreme). The provision of a symbolic peer group comes through the media, especially through the widely read women's magazines, in which domestic values are actively promoted (*Ms.* being one exception).

Peer groups are not unique to this culture. Anthropological reports often include accounts of age grades, segregated by sex, in which given activities and rituals are involved, usually oriented to adult roles. The influence of the peer group in sex role socialization is very strong, then, in that it enforces conformity to sex role standards at an individual and group level. Peer groups act as critical transmitters of societal values and induce compliance in a way that cannot be effectively resisted by most people. If David Riesman is right in his book *The Lonely Crowd,* America is at an other-directed stage of development, in which the winning of peer approval is a central mode of psychological functioning and the primary source of self-esteem for the individual. The result in terms of sex role standards is to bring the weight of society down on the individual in a very direct and effective way and to maintain the status quo of the societal sex role structure. Organized movements for sex role change may have peer impact in given instances, but their overall effect is not yet strong enough to overcome peer mediation of stereotyped expectations in most cases.

SYMBOLIC AGENTS OF SEX ROLE SOCIALIZATION

In addition to the agents of socialization that are actually present in the child's life (parents, schools, peers), he or she is touched at many points by symbolically transmitted norms of sex role standards. Prominent examples of this domain include books, toys, television, and men and women in public life (whose images are often transmitted through the media).

As has been previously mentioned, a number of analyses of

the content of children's books has revealed a dearth of female characters, while both male and female characters are often portrayed in stereotyped ways in terms of occupation and personality. Boys are often shown in adventuresome roles, girls in safe, caretaking ones. Women are almost never portrayed as working outside the home. Elizabeth Fisher[53] gives two examples of common themes. One book, *A Tree is Nice* by Janice Udry, portrayed boys climbing trees, fishing, raking leaves, etc., while the few girls in the book were seen as watching male activities or helping small children. Although not specifically *about* sex roles, much is conveyed about them. The other example, in the inanimate realm, appears in the *Best Word Book Ever* by the well-known author Richard Scarry. Male animals illustrate the meaning of activities like playing with castles, rocking horses, soldiers, and electric trains, while the (two) female animals that appear are pictured with dolls and tea sets. For the older child, biographies of prominent men abound, but there are few on women who have achieved distinction outside of traditional realms. A girl can read about the lives of Florence Nightingale and Joan of Arc, but until recently could not discover the meaning of the lives of women's rights workers, such as Susan B. Anthony.

Toys, of course, constitute a highly sex-typed domain. A report by *Ms.* magazine[54] turned up a 1972 study by Goodman and Lever in which adults were watched and interviewed at Christmastime in a toy department. They found that most buyers were quite traditional in their buying for children above the age of two, with the more creative toys (such as scientific kits) preferred for boys. Girls tended to receive dolls and other passive toys that did not make many cognitive demands or prepare for any occupational future. Boys' toys tended to be more varied and more expensive as well. Advertising for the toys tended to promulgate the sex-typed image, in that about three-quarters of chemistry sets portrayed boys on the cover, with the remainder showing both boys and girls (none showed girls alone).

Television is, of course, a rich repository of sex roles, as even a casual viewer can attest. Adventure shows most often feature males while females are relegated to ancillary and domestic positions. Women are sometimes specifically presented as sexual objects with no other real function than adornment and enhancement of the male image. Men are often portrayed in super-aggressive roles, e.g., in detective and western shows, where the moral of the story usually is that the more aggressive man wins in the end. A study by Sternglanz and Serbin[55] of television aimed at children showed a preponderance of male characters. Males were usually portrayed as aggressive and females as deferential, and boys were rewarded for their activities whereas girls received little feedback. Sternglanz and Serbin's findings also suggested that females often had to resort to magic in order to accomplish something. Even such lauded shows as *Sesame Street,* with their emphasis on liberal racial portrayals, abound in sex-stereotyped characters and actions. The amount of time children spend watching television is considerable, and so is the sex role lesson transmitted.

Women and men in public life constitute a set of role models for children (and adults as well). The most publicized women in public life at any given time usually include the wife of the President (defined in terms of his role), and various actresses, although a few female heads of state have recently joined the list. The variety of male roles represented symbolically by the media is of course much larger. Which ones capture the attention of children is another matter. Athletic heroes, adventurers (such as astronauts), and prominent political figures probably head the list, although often considerable publicity is given to other domains, such as science. Insofar as the men and women portrayed constitute a representative sample of real-world pursuits, they will almost inevitably mirror the sex role divisions and standards of the society. In acting as symbolic role models, they tend to have a symbolic modeling effect on the sex role conceptions of children.

A word might be said about the socialization effect on adults of such symbolic representations. Presumably they act to reinforce and maintain established patterns. For example, the fiction published in women's magazines (as well as the general content) advances a highly codified view of what Betty Friedan has called the "feminine mystique" (but more about that in the next chapter on the family). Advertising in particular often uses sex-typed images to sell its products.[56] For example, patent medicine commercials usually portray male authority figures curing the ills of (often) female patients. The partially clothed female image is used rather indiscriminately to sell almost anything that might conceivably be bought by men. Women in domestic roles are shown comparing the results of detergent effectiveness with a concern and concentration difficult to believe in any person above the age of six. These images are being subjected to criticism and slight revision, but they still dominate the scene. Their effect is not to create sex roles but rather to maintain and reinforce conventional conceptions of them. On almost all levels, then, real and symbolic, the child and adult are presented with an overwhelming sex role message that maintains the status quo.

THE RESULTS OF SOCIALIZATION

What is the outcome of sex role socialization? The obvious answer is, of course, men and women. What are the psychological consequences of becoming a man or woman in the full social psychological sense of the terms? That is, just by knowing a person's sex, can we predict anything meaningful about the abilities and personality traits of that individual? The answer is a qualified "yes," but before considering those areas in which sex differences do appear, we might pause a bit to look at the way psychology has conceived of masculinity and femininity as a package of traits, since this labeling process has profoundly affected the way we categorize people and think about the area of sex differences.

MASCULINITY-FEMININITY

Concern with the ideal type of man and woman considerably antedates modern psychology and personality theory. Since earliest times poets and philosophers have concerned themselves with traits that do or should characterize the ideal man or woman. In the female case, especially, there have been repeated attempts to define the "external feminine" and to appraise real women in terms of these expectations. More recently, the case of Freud is perhaps most notable in the effort to define what is inherently masculine and what is feminine in personality. Helene Deutsch's conception of woman, with its emphasis on the "feminine core" of passivity, masochism and narcissism, comes to mind as a theory with an especially clear image of the ideal woman, justified in psychoanalytic and biological terms.

An important effect of the conception of an ideal type for masculinity and femininity is that individuals are often evaluated in terms of these conceptions and that judgments of mental health or illness become based on them.[57] An especially instructive study was undertaken by Inge K. Broverman and her associates,[58] in which a sample of mental health professionals were asked to indicate their conceptions of the ideal male, female, or adult (sex unspecified). Each clinician was given only one "condition" (male, female, or adult) to rate. Results indicated that different conceptions of health existed for men and women, and that the general conception of health for "adult" more closely approximated the picture of healthy male traits rather than female ones. Therefore, a double standard of mental health for men and women exists, in which the ideal for women is perceived as being significantly different than the ideal for people in general. In the clinicians' view, healthy women were characterized as being submissive, noncompetitive, conceited about appearance, dependent, and excitable in minor crises as compared with men. The consequences of these findings can be quite serious and far-reaching for women and men. In general, professional standards for normality in mental health did not significantly differ from stereotyped conceptions of sex roles. Therefore, it is likely that men and women in therapy are being shaped to meet these standards. For example, if a male and female come in with identical complaints of unaggressiveness and lack of independence, corrective treatment is likely to be thought necessary only for the male. Yet professional standards for adult mental health in general are seen to be significantly different from those prescribed for women. Therefore, the traits conceived of as normal for women, which may be the goals of treatment, are seen as inadequate for adult functioning in general. In fact, the constellation of traits seen as constituting healthy femininity seems to be quite childlike and not consistent with adult responsibilities.

We have seen that psychologists see men and women in a way not very different from that of society in general. They have even devised tests with which to evaluate the person in terms of his or her display of these traits.[59] Generally the validity of test items is established by choosing those items that discriminate reliably between males and females. Masculinity-femininity has been seen as a unidimensional continuum defined by masculinity and femininity as poles. Thus, if a person were high in femininity, he or she would by definition be low in masculinity. It is doubtful, however, that masculinity or femininity is such a simple trait that it can be so easily measured or assessed. Thus, a set of behaviors and attitudes may characterize the ideal type of masculinity or femininity, but they may be multidimensional in structure, not additive. In addition, a person might possess traits assessable on both scales, masculine and feminine, and not simply by one or the other.

Opinions have differed on the usefulness of the conception itself. An early and influential study by Terman and Miles[60] saw "mental masculinity and femininity" as core traits of temperament. They devised a test in which items were selected on the basis of divergent performance by males and females; thus, one response to an item was scored as feminine, the opposite as masculine. Other similar tests are the masculinity-femininity scales of the Strong Vocational Interest Blank and the Minnesota Multiphasic Personality Inventory (MMPI). Both of these tests are widely used in counseling and clinical settings. The use of such tests can often have unfortunate effects, in that deviance from sex stereotypes is defined as aberrant behavior and is sometimes seen as connoting sexual inversion. A woman who has interests that differ from those of most other women taking the Strong test will be characterized as "masculine," a term of opprobrium perhaps only slightly less damning than "feminine" used to categorize a male. This labeling process can have bad effects on the person taking the test as well as on the views of clinicians and counselors using such tests. An unnecessarily rigid, either-or concept of sex roles is implied by the masculinity-femininity conception.

SEX DIFFERENCES IN COGNITIVE ABILITIES

Before plunging into the controversial area of sex differences in cognitive abilities. let us caution the reader against overinterpreting these differences. In all cases, the finding of a "sex difference" implies the existence of a *mean* difference, i.e., when the average scores of women and of men are compared, a difference can be discerned. It does *not* mean that every man is different from every woman; in fact in most cases there is a large overlap (so that most men and most women score within the same range), but taken overall, a difference can be found. It is also easy to interpret such a "difference" as being largely responsible for sex differences in career choice. For example, the sex difference in spatial ability has been used as an explanation for the low participation of women in the sciences. This is definitely an overinterpretation of the findings. Ability is only one factor in career choice, and a relatively large range of abilities is encompassed within any one career category. It is the author's contention, as will become apparent in the chapter on the family system, that most of the variance in career choice can be accounted for by motivational factors stemming out of differential childhood socialization and the nature of the family achievement system. These very factors often influence the outcome of the ability tests themselves, as we shall soon see.

Because of the scientific paradigm, which defines a positive result in terms of a difference found, and also because of the popular conception of the divergent nature of men and women, the effects of sex on personality have been conceived of as sex *differences.* Researchers tend to forget areas in which no differences have been shown and concentrate their study in areas that do show such differences, thereby biasing the literature and psychology's conception of men and women. With this phenomenon in mind, we will consider sex differences in cognitive abilities and temperament, saving our discussion of achievement motivation for the chapter on the family system, since such differences and their socialization are so intimately connected with it.

The three areas chosen for discussion, spatial abilities, verbal abilities, and creativity, have been singled out because they are most often cited for their relevance to real-life sex differences in career participation. Before considering each of these, we shall pause a bit to consider the general issue of sex differences in intellectual functioning. One persistent issue in the intelligence testing controversy, which has been highlighted by recent dissension over racial differences, is the validity of such testing and its meaning outside of the testing context. No short discussion can do justice to all the complex issues. The selection of test items has been criticized as being culturally and sexually biased. There is some truth to this in the case of sex differences, although perhaps the bias has not been as great as in the case of race. Boys and girls do to a limited degree live in different cultural worlds. Perhaps more important, they learn society's expectations about the sex role evaluation of those worlds. So if the typical mathematical problem is set up in terms of rocket ships or race cars (as it often is), a signal is sent to the girl of a mismatch between her sex role and the content of the problem. This mismatch may lead her to expect herself to perform more poorly and not to try as hard, motivational factors which will receive fuller consideration in the next chapter. Studies have been done in which female sex role items were substituted for male in math problems, and female performance did improve. A more general issue in the area of intelligence testing is the relevance of test performance to real-world performance. Aptitude test scores do seem to relate to school grades, but school grades seem to have a somewhat tenuous link to real-world performance, probably because so many motivational and social-structural variables intervene to affect achievement. Therefore, in interpreting test scores we must remember that they are only one possible indicator of future achievement.

Eleanor Maccoby[61] has speculated on some general causes for sex differences in intellectual functioning. Her theory is that optimum intellectual performance is facilitated by a moderate amount of boldness and impulsivity. The distribution of the male population favors the high end of the scale, with high degrees of boldness and impulsivity having a negative impact on performance. On the other hand, it is relatively rare for girls to be high in boldness and impulsivity, and when they deviate from the optimum it is in the direction of passivity and inhibition, which inhibit good performance at intellectual tasks. Therefore, Maccoby asserts, the factors that impede intellectual performance may be different ones for boys and girls, depending on the population distribution of the traits for the two sexes. A girl, then, who is a little more bold and impulsive than average for her sex might be expected to excel, whereas a boy with a lower level of these attributes than average for males might be a high scorer. Corroborating evidence comes from correlation studies that show that impulsiveness correlates positively with performance on analytic tasks for girls but negatively for boys. Acting entirely in accordance with sex role seems to be a negative factor for performance in both boys and girls. Neither "real" boys nor "real" girls, defined in terms of ideal types of sex role standards, seem to excel at such tasks. A bit of deviance for both seems to be a good thing as far as intellectual functioning is concerned.

Before going on to consider the three specific areas of interest, let us discuss two rather controversial theories that attempt to explain the origin of sex differences in cognitive abilities. The first theory is offered by Broverman, Klaiber, Kobayashi, and Vogel[62] and concerns the roles of activation and inhibition in sex differences in cognitive abilities. They claim to explain female superiority in simple perceptual motor tasks and male superiority in more complex tasks involving stimulus reorganization by recourse to sex differences in the balance between the adrenergic activating and the cholinergic inhibitory neural processes. According to these investigators, the sex hormones act differentially on these systems to account for the effect. Estrogen is said to favor greater activation through the adrenergic or sympathetic nervous system. This system is said to be involved in facilitating performance on simple perceptual motor tasks that depend on prior experience and fine coordination rather than on the creation of novel responses or integration of new material. Estrogen is also said to decrease the inhibiting direction of the cholinergic or parasympathetic nervous system. The cholinergic system is an inhibiting system that acts to delay initial responses until a higher integration is effected with the production of novel responses. Tasks at which males excel tend to involve such factors. Androgen also seems to act as an activating agent, however, but the authors suggest that estrogen is more potent in this regard, thus altering the activation-inhibition balance in favor of greater activation for females. This behavioral activation leads to a lesser ability to inhibit responses in favor of higher integrative functions, resulting in a deficit in such information-processing skills.

This theory has been criticized on several points. A critique by Singer and Montgomery[63] challenged the notion of the adrenergic-cholinergic division on neurophysiological grounds, as well as some of the experimental evidence on which the hypothesis is based. For example, in the original Broverman et al. article, a number of rat studies were cited as evidence for the behavioral activation of certain responses, whereas the critique cites studies in which cholinergic (inhibitory) facilitation of the same responses was noted. A second critique, by Mary Parlee,[64] charged that many of the behaviors cited in the original study were "at best tangentially related to human cognitive abilities and rely for their relevance on some dubious cross-species analogies."[65] In addition, she regarded the selection of studies as biased and the relationship of the sex hormones to the activation-inhibition balance as unproved. The Broverman et al. theory, then, remains a provocative and original one, but it must definitely be relegated to the "as yet unsubstantiated" category.

Another shotgun-theory attempt to explain sex differences in cognitive abilities was presented by the British psychologists Jeffrey A. Gray and Anthony W. H. Buffery.[66] They maintain that the adaptive significance of spatial and verbal abilities account for their differential distribution between the sexes. Aggression is said to relate to the organization of dominance hierarchies, the maintenance of which is facilitated by greater spatial ability in the male. Spatial ability is also necessary for successful defense of the group. Given the sexual division of labor called for by reproductive functions, social organization and defense are given over almost entirely to the male since the female is occupied with the care of the young. Greater female fearfulness in primates is noted as a facilitator in the organization of dominance hierarchies, which act to stabilize the group. According to this theory, females also display more developed verbal skills due to the adaptive advantage of a verbally rich mother-infant environment. The neural areas for fearfulness and verbal ability appear to be linked; therefore it is possible that selective pressure favoring the development of one of these traits would bring the other along as an automatic secondary consequence. Therefore, greater female fearfulness could be a result of natural selection for high verbal ability as well as a factor in facilitating dominance structures. It has been suggested that endocrine secretions support this system, with androgen favoring the expression of aggression and estrogen inhibiting fear, but this goes against the direction of the theory. The latter result is complicated by the fact that in rodents, males are more fearful; in primates, the opposite is the case. The difficulty in explaining fearfulness through the action of hormones, as well as problems in conceiving of direct hormonal effects on entire behavioral systems (fearfulness, aggression, etc.) have led to a critique of this work by John Archer.[67] Gray and Buffery's theory, then, has not enjoyed great acceptance among psychologists. That sex differences in cognitive abilities are at least partially adaptive seems plausible, but no coherent picture of such effects has yet been presented.

We will now specifically consider the three areas of cognitive abilities of special concern to sex role researchers: spatial abilities, verbal abilities, and creativity.

Spatial Abilities

Probably the most commented-upon sex difference in spatial abilities is in psychological differentiation, better known as field dependence-independence. This concept, developed by Herbert Witkin and his associates,[68] involves the ability to separate distinctive features from a competing background, or in other words to ignore some contextual input and instead to identify one feature of interest. Two of the best-known measures of this concept are the rod and frame test and the embedded figures test. In the rod and frame test, the subject must adjust a rod to the perpendicular in a field of competing spatial inputs, and in the embedded figures test, hidden irregular geometric figures are to be found in a jumble of background material. Sex differences in field dependence do not emerge until adolescence and then remain throughout adulthood.[69] Females tend to be field-dependent, i.e., unable to adjust for the competing contextual inputs, and males tend to be field-independent, i.e., they score better on those tasks that involve separating one element from a field. Such differences, although small, have been interpreted as indicating the relative superiority of males in the "analytic" style of thought, presumably a prerequisite for scientific skill. David McClelland has put the comparison less pejoratively: "Women are concerned with the context, men are forever trying to ignore it for the sake of something they can abstract from it."[70]

But do such small differences on relatively specific tasks justify the types of generalizations about sex differences that have been drawn from them? Field dependence does correlate with measures of interest in people and interpersonal situations, leading to the inviting conclusion that global attributes of maleness and femaleness are all somehow grounded in differing cognitive styles or that all attributes, cognitive and socio-emotional, somehow derive from a common organizing principle of maleness or femaleness. The effort to find such principles is relatively common within psychology. For example, Silverman[71] advocates the existence of masculine and feminine principles in thought, which he calls Logos and Eros. Logos is characterized as being objective, analytic, rational, independent, and emotionally controlled, while Eros is seen as intuitive, affective, and oriented to wholeness and relatedness rather than to separation and analysis. The parallel to the field-dependence work is easy to see. Likewise, Ravenna Helson[72] has developed the similar notions of patriarchal and matriarchal consciousness, derived from Neumann and ultimately from Carl Jung. She analyzes work by male and female mathematicians and authors to substantiate her theory. David Gutmann[73] has maintained that allocentric and autocentric ego styles differentiate males from females. The male allocentric style is basically similar to the previously mentioned notions of Logos and patriarchal consciousness, with the specification that the feeling of separation between self and the world of objects and persons is the basis for such a style. The female autocentric style implies a lack of boundary between self and the external world, resulting in qualities akin to Eros and matriarchal consciousness.

Although most researchers in the field-dependence area are careful in generalizing the results of their research, it does tend to get assimilated to the above-mentioned tradition of assuming wide disparities between male and female experience and cognitive processing of the world. This tradition is of course similar to that behind masculinity-femininity testing, mentioned earlier, although it differs in that it advocates a unitary organizing principle rather than a string of related behaviors. In a recent extensive review of the literature, Maccoby and Jacklin[74] find few consistent sex differences in general analytic skills, which presumably should be related to the field-dependence tests if the global hypotheses are correct.

Explanations for the difference in field dependence abound, although sometimes one wonders if all the theorizing is justified by the small magnitude of the differences involved. The general hypothesis of Broverman et al. on activation and inhibition has already been presented, and we have seen that little undisputed evidence exists for its validity. An intriguing hypothesis on the heritability of sex-linked genes affecting spatial abilities has been presented.[75] A number of studies have demonstrated cross-sex correlations in spatial ability between parents and children, i.e., father-daughter and mother-son correlations, but not same-sex parent-child correlations. Such a pattern would be consistent with the explanation of a recessive gene for spatial ability carried on the X chromosome. If such were the case, then approximately twice as many males as females should manifest the trait. However, this gene (if it exists) would not be the sole determiner of spatial abilities; even on a genetic level, other genes would be implicated. The issues in genetic transmission of such a general trait are quite complex, and many geneticists refute the possibility of any central role for heredity at all. And, of course, even if genetic factors are operative, cultural enhancement of such effects is not ruled out. If we are ever able to apportion variance for the spatial sex differences effect accurately, genetic factors may turn out to play an ancillary role to the primary one of differential training and sex role expectations.

Much of the thinking about spatial abilities has concerned the effects of differential experience and sex stereotypes as primary shapers of such abilities. As we have seen, boys and girls are given very different toys to play with starting at an early age. Boys are given trucks and blocks and erector sets, all of which develop spatial skills, while girls entertain themselves with dolls and tea sets. Role models for scientific achievement are scarce for girls; even advertisements for chemistry sets usually picture only boys. The generally inculcated traits of independence and dependence may generalize to performance in the spatial domain. In this area, of course, girls would be at a disadvantage on the field-dependence tasks since their training emphasizes relatedness to others and dependence on them. Still more difficult to pinpoint may be the subtle but devastating effects on a girl of knowing that her life plan is unlikely to include participation in the analytic, scientific domain. Her expectations of herself and motivation to succeed may therefore be inhibiting factors in her performance. She may even learn to attach negative labels to attempts to excel in such male-dominated domains, so that an emotional effect disrupts her performance.

In any case, our real task is to assess the significance of such sex differences for real-life problems of occupational choice and achievement. In examining this question, we might want to take a look at abilities that are more specifically relevant to real-life choices. Mathematical ability, which is linked to spatial and analytic skills, is the basis for entry into any one of a large number of positions in scientific and technical endeavor. These occupational areas, of course, are heavily dominated by men (at least in this society, and to a lesser degree in others as well, e.g., China and Russia). Tests of mathematical skills (e.g., the College Board mathematical aptitude test) consistently show male superiority, and in the realm of professional mathematics, males form a preponderant majority (when the largely female secondary-schoolteaching segment is eliminated). In the census of truly creative mathematicians, women constitute a pitiful minority (Helson[76] was able to find eighteen, as designated through peer nomination). Even if mean sex differences in spatial and mathematical ability are small and the distributions largely overlap, the upper end, the likely source of truly unusual talent, is heavily biased in the male direction. Of course, it is entirely possible that the same environmental factors that go into career choice affect performance on tests of mathematical skills as well.

Lynn M. Osen,[77] a biographer of seven prominent women mathematicians, has coined the term "the feminine mathique" to denote the constellation of cultural values and beliefs that surround women and mathematics. According to Osen, the feminine mathique "breeds and institutionalizes graceless jokes and stereotypes about the helpless, checkbook-bumbling female, the mindless housewife, the empty-headed husband-chasing coed, the intuitive (but illogical) woman who 'hates arithmetic.'"[78] Achievement in mathematics is often considered unfeminine, even if it does emerge despite all internal and external obstacles. Psychological thinking about the divisions of Logos and Eros and patriarchal and matriarchal consciousness almost exactly mimes the thinking of most men and women when it comes to the mode of thought seen to be appropriate to each sex. Probably the most potent factor in discouraging women from mathematical and scientific careers is the lack of motivation for any sort of sustained commitment to an occupational endeavor at all, especially if it involves achievement in a male-identified field. If women are socialized almost from birth to think in terms of a life plan that does *not* include a major career commitment, but rather emphasizes marriage and family responsibilities, the chance of discovering and using their mathematical or scientific talent approaches zero. For the male, the decision is not whether to choose a career, but which career to choose. Much more attention will be paid to this issue in the next chaper, since the nature of the family achievement system is at the root of the socialization process. This issue will be doubly relevant to the section that follows, which discusses verbal abilities, since this is an area in which women generally outperform men, yet the number of women pursuing professional writing careers is small.

Motivational and social structural factors, then, define the areas of appropriate achievement for men and women; crossing these barriers is exceedingly difficult for psychological and practical reasons. Nowadays there is considerably more opportunity for women in the sciences and mathematics than formerly, although outright discrimination and negative attitudes on the part of male colleagues are still relatively common. Until quite recently, the "proper" education for a woman definitely did not include exposure to higher mathematics, and this doubtless depressed the number of women who were able to discover their talent, let alone use it. Osen gives a rather dramatic example in the case of Mary Fairfax Somerville (1780-1872) who discovered some algebraic symbols in a fashion magazine (of all places) that she read at a tea party (ditto). Her curiosity was stimulated, and she embarked on an ambitious and solitary course of study, which resulted in her publication of a number of highly regarded treatises in the natural and physical sciences. She studied despite the disapproval of her family; in fact, her mother actively intervened by confiscating the candles she was using to read Euclid's geometry. Finally she was able to acquire a number of mathematical works, and in a memoir she has described the difficult path of her quest for knowledge:

> I was thirty-three years of age when I bought this excellent little library. I could hardly believe that I possessed such a treasure when I looked back on the day that I first saw the mysterious word "Algebra," and the long years in which I persevered almost without hope. It taught me never to despair. I had now the means, and pursued my studies with increased assiduity; concealment was no longer possible nor was it attempted. I was considered eccentric and foolish, and my conduct was highly disapproved by many, especially by some members of my own family. They expected me to entertain and keep a gay house for them, and in that they were disappointed. As I was quite independent, I did not care for their criticism. A great part of the day I was occupied with my children; in the evening I worked.[79]

The societal attitudes and socialization pressures experienced by Mary Fairfax Somerville still exist today and inhibit the intellectual development of untold numbers of talented women. The contribution of genetic and biological components to performance on spatial tasks and mathematical tests is uncertain, but the overriding influence of socialization and societal constraints is incontrovertible. The viability of global conceptions of masculinity and femininity has not been established, except for the fact that they are an almost exact mirror of societal values and beliefs about sex roles. Therefore we close this section emphasizing the dominant influence of social factors, a bias that will continue in the next section when we consider the dilemma of superior female verbal skills and inferior female verbal achievement.

Verbal Abilities

Maccoby and Jacklin, in their mammoth study of sex differences, assert that "female superiority on verbal tasks has been one of the more solidly established generalizations in the field of sex differences."[80] Language tends to emerge earlier in females, although some recent work disputes this finding. Up to the age of ten or eleven, consistent sex differences are rare, but after this age female superiority emerges. This difference remains through adulthood, but again, as with the spatial ability, it is not large. Throughout the school years, boys lead girls in reading problems and in speech difficulties, some of which might be a result of the greater disciplinary problem in compelling boys' attention to school tasks.

Early sex differences in language development have been explained in a number of ways, including greater mother-daughter verbal interaction, which does seem to be supported in a number of studies. Greater physical activity and more time away from the mother among males might be partially responsible for the discrepancy rather than conscious maternal attention to female language development. Another possibility for explaining girls' verbal achievement lies in their probably greater exposure to reading. As a quiet, sedentary, "ladylike" activity, reading is considered an appropriate pastime for girls and may even be encouraged instead of the rough outdoor play more typical of boys. Boys may stay at home less and be less likely to sit still for the extended period of time necessary for reading. Parents (especially mothers) may also be more likely to read to daughters than sons because of their presence indoors and because of the sex-appropriateness of this sort of quiescent activity.

Of course, other more controversial theories of female verbal superiority have been offered to compete with the almost prosaic one of differences in socialization. One of the most intriguing theories involves the issue of brain lateralization of abilities.[81] Experience with stroke victims and other cases has shown that the left hemisphere of the brain coordinates speech and verbal abilities, whereas the right hemisphere is largely involved in the processing of nonverbal material, including spatial information. Some evidence has been offered to support the theory that girls develop left-hemisphere dominance earlier, thus facilitating verbal skills. Females seem to show greater dominance (or laterality) than males, and Buffery and Gray hypothesize that this inhibits the development of spatial skills. This problem is exceedingly complex, however, and competing explanations and discordant findings abound. For example, as Maccoby and Jacklin point out, there is weak correspondence between sex and the entire "package" of skills attributed to each hemisphere. Thus, they cite Levy-Agresti's[82] description of the left hemisphere as "verbal, sequentially detailed, analytic and computer like,"[83] and comment that the last three traits do not accord with common "female" qualities. Too little is known about the specific operation of the brain hemispheres and their interactions to support a definitive view of their place (if any) in accounting for sex differences in verbal and spatial abilities.

Given female superiority in tests of verbal abilities, why do we not have more great (or even not so great) women writers? Using the same "logic" that has been applied in the case of spatial abilities, we should expect verbal ability to be the major factor in career choice and achievement. Thus, the usual explanation for male dominance in mathematics and science involves a central role for differences in spatial and mathematical ability, with a perfunctory nod to social factors. Yet in the case of verbal abilities we have a complete reversal: higher female verbal abilities, yet lower female participation in relevant occupations. Granted that the number of truly creative women writers far exceeds that of similarly placed women mathematicians and scientists, and that large numbers of women are found in ancillary writing posts, when we come to the highest-level jobs in publishing and writing, women are still a minority. Of course, it can be argued that the skills measured in verbal tests do not exactly match those needed for creative and skillful writing, but there certainly must be some correlation. (For that matter, the spatial skills measured on the embedded figures test do not exactly parallel the skills needed to devise a theory of relativity either.) In this instance it is clear that social factors intervene to separate high verbal ability from exercise of that ability in a career. All of the socialization and social structural factors previously mentioned pertain to this case, in that choice of a career, any career, is still an anomaly for women and a *choice* that can be forsaken in favor of the less demanding path of a domestic existence. Again, more extensive treatment of this subject will be found in the next chapter.

Perhaps the best evocation of the plight of the female writer has been given by Virginia Woolf in *A Room of One's Own.* In a vignette of a hypothetical sister of Shakespeare who shares his talents and attempts to use them, Woolf portrays the force of social circumstance that has kept women in their place:

She was as adventurous, as imaginative, as agog to see the world as he was. But she was not sent to school. She had no chance of learning grammar and logic, let alone of reading Horace and Virgil. She picked up a book now and then, one of her brother's perhaps, and read a few pages. But then her parents came in and told her to mend the stockings or mind the stew and not moon about with books and papers. They would have spoken sharply but kindly, for they were substantial people who knew the conditions of life for a woman and loved their daughter—indeed, more likely than not she was the apple of her father's eye. Perhaps she scribbled some pages up in an apple loft on the sly, but she was careful to hide them or set fire to them. Soon, however, before she was out of her teens, she was to be betrothed to the son of a neighbouring wool-stapler. She cried out that marriage was hateful to her, and for that she was severely beaten by her father. Then he ceased to scold her. He begged her instead not to hurt him, not to shame him in this matter of marriage. He would give her a chain of beads or a fine petticoat, he said; and there were tears in his eyes. How could she disobey him? How could she break his heart? The force of her own gift alone drove her to it. She made up a small parcel of her belongings, let herself down by a rope one summer's night and took the road to London. She was not seventeen. The birds that sang in the hedge were not more musical than she was. She had the quickest fancy, a gift like her brother's, for the tune of words. Like him, she had a taste for the theater. She stood at the stage door; she wanted to act, she said. Men laughed in her face. The manager—a fat, loose-lipped man—guffawed. He bellowed something about poodles dancing and women acting—no woman, he said, could possibly be an actress. He hinted—you can imagine what. She could get no training in her craft. Could she even seek her dinner in a tavern or roam the streets at midnight? Yet her genius was for fiction and lusted to feed abundantly upon the lives of men and women and the study of their ways. At last—for she was very young, oddly like Shakespeare the poet in her face, with the same grey eyes and rounded brows—at last Nick Greene the actor-manager took pity on her; she found herself with child by that gentleman and so—who shall measure the heat and violence of the poet's heart when caught and tangled in a woman's body?—killed herself one winter's night and lies buried at some cross-roads where the omnibuses now stop outside the Elephant and Castle.[84]

With suitable changes for the passage of time and the demands of writing careers other than that of the dramatist, the outlines of the malevolent and benevolent forces of socialization and social pressures remain unchanged. Even if a woman should overcome the formidable obstacles to achievement, her work may often lie forgotton or be maligned by male critics as reflecting the inferior sensibility of the "female mind."

Another issue is important in the discussion of female achievement in literature (and in mathematics and science as well), and that is the issue of creativity. Are men more creative than women? Our next section will examine this question.

Creativity

When asked to name the most creative children in their classes, teachers named more boys than girls. However, when more exact measures of creativity have been administered, the results are much more complex. Although no consistent differences have been demonstrated on measures of divergent thinking, sex differences in correlates of creative performance abound. Evidence has shown that creative men (such as architects) achieve more "feminine" scores on conventional tests of masculinity-femininity. A study on boys also reported higher creativity scores in those boys who showed a mixture of male and female sex-typed interests. Some sex role deviance then seems to correlate with high levels of masculine creativity. However, the complementary hypothesis that girls who have more "masculine" interests will be more creative has not been supported. High defensiveness has been shown to be a creativity inhibitor in boys but not (to such an extent) in girls. Kogan[85] has suggested that defensive boys keep quiet in threatening situations (such as a creativity test) and hence do not produce as many verbal responses, while the more verbally socialized girls do not respond in this way (or may actually increase their verbal output, but not the uniqueness of their ideas). The typical divergent thinking task involves the production of ideas for the use of a common object (e.g., a brick), with ideas scored for number and uniqueness, so that any inhibition of verbal behavior would lower scores.

The real question, of course, is not comparative scores on creativity tests but real-world accomplishment. Since all creative endeavor occurs in some constraining social context, it is necessary to understand the nature of that context before jumping to conclusions about the origin of differences. When Nobel Prizes are counted, for example, the sex composition of the scientific pool must be considered. The real inhibitions on creativity seem to lie in the structures that channel men into potentially creative occupational fields and women away from them and into the home. The next chapter, on the family, considers these influences in depth. In our discussions of the mathematician Mary Fairfax Somerville and of Virginia Woolf's imaginary sister to Shakespeare, we have already considered some of these issues. Even when women are drawn into potentially creative occupational fields, such as the sciences, socialization and societal pressures may draw them away from research and toward teaching, a more "feminine" endeavor. Studies of scholarly productivity, however, have not found differences in publication rates for women and men, although the quality of such works was not assessed. In peer nomination studies, men far outnumber women in being labeled as "creative," but we have to consider all the factors that go into selection for the occupation itself before we can understand this phenomenon. "Creativity" is only the final product of a lifelong series of pushes and pulls that go into shaping one's life according to a sex role pattern.

SEX DIFFERENCES IN SOCIOEMOTIONAL TRAITS

Developmental studies of sex differences in socioemotional behaviors have not shown the sorts of differences one might expect from knoweldge of the adult stereotypes of the two sexes. In an exhaustive review of the literature, Maccoby and Jacklin[86] report few or no consistent childhood sex differences in the following areas: emotionality, timidity, dependency, social responsiveness, nurturance, and altruism. Studies often showed no differences or differences that were highly dependent on the age group or the specific situation involved. Adult sex differences in such behaviors, then, might be more a function of differential opportunity to express such traits rather than of any innate or early learned predisposition.

Differences were found in activity and aggression, with boys exhibiting more of both behaviors. In the study of aggression, observational studies, modeling studies, dream records, projective tests, and other measures all reported the same result: more aggression in males. Females were less often involved in naturally occurring aggression as either instigators or victims. In staged studies of shock administration, females were given less shock. In Chpater 1, we reviewed the evidence for a hormonal stimulus for male aggressive behavior. In the next chapter we shall see that the structure of the family and of the work world also direct the socialization of the male in a more aggressive direction. Although excessive aggression is viewed as a problem for boys, so is deficient agression, which is not considered a problem for girls. Training for independence and achievement draws from this base of male aggression, as we shall see in the next chapter.

Girls and women seem to be more compliant and subject to external influence. In the younger years, girls are more likely to comply with an adult request than boys are, indicating that they may have been socialized more for adult approval than boys are. One theory about this holds that the "feminine" traits of beauty and pleasing personality depend on social consensus for their reality, while "masculine" traits are more often based in a more impersonal definition of ability and achievement. Therefore, if a girl is trained to display the traits of conventional femininity, she will as a matter of course be more dependent on and responsive to the opinions of others.

The Parsons-Bales[87] theory of instrumental and expressive distinctions parallels the findings of the psychological studies. In a small group, two processes are necessary for smooth group functioning: (1) instrumental orientation toward the achievement of a goal and (2) a socioemotional outlet for expression of emotion and support. Studies of such small groups tend to confirm the existence of an instrumental-expressive split in functioning, with males usually predominating in the first area and females in the latter. Parsons and Bales have also extended their theory to explain the functioning of the family, with the father representing the instrumental, and the mother the expressive, component. Male predominance in aggression, dominance, and achievement would correspond to the instrumental orientation, while the female tendency to compliance and persuasibility would reflect an expressive or socioemotional tendency. Although consistent differences in childhood social responsiveness have not been found, it is clear that by adulthood females operate more often in the interpersonal domain and males in the object-oriented, instrumental one. In a study of college-age male and female same-sex groups, Aries[88] found females much more likely to talk about themselves and their personal relationships, while males were most often concerned with themes of competition and status. The relative lack of childhood sex differences in social behavior would argue for the operation of adult channeling of interests and exposure to sex-typed environments as important components of socialization.

Adult sex differences in personality and interests can best be understood as the result of the twin influences of childhood socialization pressures and adult channeling. An often-cited cross-cultural survey of sex differences in socialization[89] found that ethnographic reports of childhood socialization in 110 mostly preliterate cultures emphasized sex differences in four major areas: nurturance, responsibility, achievement, and self-reliance. Girls were most often trained for nurturance and responsiblity, boys for achievement and self-reliance. These differences are, of course, intimately tied to the structure of the family and the occupational achievement system. Socialization pressures may only show their full effect when adult sex role environments channel behavior in consistent ways. Much recent sex role research has dealt with achievement motivation and independence training, subjects we shall save for the next chapter, since they so closely relate to adult sex differences in occupational participation.

In a survey of the ideology of the feminine character in psychological theory, Viola Klein reported that the traits most often associated with women were "passivity, emotionality, lack of abstract interests, greater intensity of personal relationships, and an instinctive tenderness for babies."[90] We have already considered some of these issues in our discussions of the psychoanalytic perspective, the menstrual cycle, and maternal behavior, and shall not repeat them here. The nature of one's everyday activities and life plan will of course shape the sort of personality traits one exhibits, as we shall see in our chapter on the family. Societal images of the ideal feminine type, or what Betty Friedan has called the "feminine mystique" also affect the expression of traits.

The study of the masculine character as a separate area of inquiry is a rather recent innovation. While extensive attempts have been made to identify the essence of femaleness, the male character has more or less been assumed to reflect the natural state of "humanity." To use Simone de Beauvoir's terms, man is the standard, and woman is "the other." Recent movements in sex role reform for men have concerned themselves with defining masculinity as a specific role prescription of personality traits rather than a general description of human nature. Marc Feigen Fasteau has described what he calls "the male machine":

> The male machine is a special kind of being, different from women, children, and men who don't measure up. He is functional, designed mainly for work. He is programed to tackle jobs, override obstacles, attack problems, overcome difficulties, and always seize the offensive. . . . His circuits are never scrambled or overrun by irrelevant personal signals. . . . His relationship to

> other male machines is one of respect but not intimacy; it is difficult for him to connect his internal circuits to those of others. In fact, his internal circuitry is something of a mystery to him and is maintained primarily by humans of the opposite sex.[91]

In any discussion of the male character, themes of aggression and competitive achievement are dominant, with emotional control and social insensitivity secondary considerations. Sex differences in aggression, biologically primed, may account for many of these differences as well as for the societal (male) structures of competitive achievement that promote and reward the display of masculine character traits. Of course, such structures can become functionally autonomous of whatever biological base may have engendered them, and it is likely that this has been the case. Societal glorification of the military ideal and of aggressive power is only the tip of the iceberg of the idealization of masculine traits. High occupational achievement (especially when accompanied with high salary) has been adopted by many women as a final goal for defining female liberation (although some have disputed the wisdom of adopting such a male model). The operation of the achievement-occupational-domestic system is so important to our understanding of sex differences in socioemotional traits that we will devote the entire next chapter to it.

SUGGESTED READINGS

Phyllis Chesler. *Women and Madness*. New York: Avon Books, 1972 (p). How women are dealt with by the mental health establishment: a feminist view.

James S. Coleman. *The Adolescent Society*. New York: Free Press, 1961 (p). The classic sociological study of adolescence.

Violet Franks and Vasanti Burtle, eds. *Women in Therapy: New Psychotherapies for a Changing Society*. New York: Brunner/Mazel, 1974 (p). Deals with various therapeutic techniques and the special psychological problems of women.

Viola Klein. *The Feminine Character: History of an Ideology*. Urbana: University of Illinois Press, 1975 (p). How psychologists (and others) have viewed femininity.

Gisele Konopka. *Young Girls: A Portrait of Adolescence*. Englewood Cliffs, N.J.: Prentice-Hall, 1976 (p). Reports on interviews with adolescents, showing their conflicts about sex roles.

Eleanor E. Maccoby, ed. *The Development of Sex Differences*. Stanford: Stanford University Press, 1966. Well-known and often-cited set of theoretical pieces discussing psychological research. Comprehensive annotated bibliography.

Eleanor E. Maccoby and Carol N. Jacklin. *The Psychology of Sex Differences*. Stanford: Stanford University Press, 1974. Best book in the area; summarizes the state of research and includes a comprehensive annotated bibliography.

Jean B. Miller, ed. *Psychoanalysis and Women*. Baltimore: Penguin Books, 1973 (p). Good collection of classic and contemporary psychoanalytic perspectives on women, modifying Freud's original views.

Jean B. Miller. *Toward a New Psychology of Women*. Boston: Beacon Press, 1975. Provocative and promising ideas on revising the psychological image of woman.

Juliet Mitchell. *Psychoanalysis and Feminism: Freud, Reich, Laing and Women*. New York: Random House, 1974 (p). Overview and critique of Freud's work. Sympathetic but feminist. Some coverage of other theorists as well.

Julia A. Sherman. *On the Psychology of Women: A Survey of Empirical Sources*. Springfield, Ill.: Charles C. Thomas, 1971. Good summaries of research areas in the psychology of women. A fine place to start seeking other sources.

Judith Stacey, Susan Bereaud, and Joan Daniels, eds. *And Jill Came Tumbling After: Sexism in American Education*. New York: Dell Publishing

Co., 1974 (p). Collection of articles discussing issues like school texts, teacher attitudes, and so on.

Jean Strouse, ed. *Women and Analysis: Dialogues on Psychoanalytic Views of Femininity*. New York: Viking Press, 1974 (p). Excellent anthology of primary sources with new essays by Margaret Mead, Erik Erikson, and others.

3

The Social Maintenance System: The Family

The family system intersects the sex role system at two critical points: the family of orientation and the family of procreation. One's family of orientation includes parents and siblings, as defined by family of birth, while one's family of procreation comes into being at marriage and includes spouse and children. The two systems, family of orientation and family of procreation, are not independent, of course. Normally, the structure of roles, particularly sex roles, remains constant across both and provides a link of historical continuity across the generations that defines the nature and stability of a society. In addition, the socialization one receives in one's family of orientation is normally largely preparatory for participation in one's future family of procreation.

The socialization processes of identification, imitation, and social learning take place in the family of orientation. Children are born into a given family structure, usually that which is typical for the society (in ours, the nuclear family) and see and emulate adult models of their sex. They gradually learn that what society is largely about is just this chain of generations, and their family socialization is designed to impress this upon youngsters. In contemporary American society, there has been a turn away from socialization for family participation to socialization for individuality and achievement, but even the latter is most often integrated into the family structure. The more traditional the society, the more likely it is that one's individual destiny will be entwined with the patterning of family units. Even in our divorce-ridden society, however, traditional family values remain strong for a large part of the population and define the aspirations of most people, even if not always their achievements.

The family of orientation, then, remains the prepotent site for early sex role socialization and gives the child the closest and most prolonged contact with enactors of adult sex roles, thus providing a picture of adult life that is likely to be a lifelong influence on the child. The last chapter reviewed research on the processes of parental sex role socialization, so we shall not repeat the discussion, except to note that much of that socialization is concerned with preparation for adult *family* roles. The point is perhaps obvious in the case of the twentieth-century American female, with domestic values, play, and aspirations being an early and integral part of sex role socialization. The case of male socialization is perhaps less clear, since we so often tend to confuse male aspirations with human endeavor in general. Boys are taught early the competitive, achievement-oriented values that define male performance in this society. They are also given notice of the need to define an occupational identity, which, although it may or may not be personally satisfying, at the very least must provide economic support for the male's future family of procreation. We may tend to glorify this requirement by calling it "opportunity," but it is often perceived as a very real pressure by males. For the most part, in our society a male is defined occupationally, and his worth is judged by his prestige, education, position, and the money he earns. These factors are often reflected in the manner of living of his family of procreation, which highlights the importance of the economic support function of the male sex role.

We shall wait until the end of the chapter to discuss the nature of socialization for family roles, in particular achievement motivation, fear of success, and the so-called "feminine mystique" of domestic values. Our point here is that understanding these processes requires a prior comprehension of the nature of the structure of the family itself. For this, we shall have to make a brief excursion into history (and historical speculation) and economics, as well as into political structure. The purpose of this journey is to examine more closely the impact of the family of procreation on our adult sex role lives. At every age, most men and women are largely constrained in their roles by the limits of this system. Understanding the possibilities for sex role change also requires knowledge of the family system, since it has most often been this factor that has impeded attempts at sex role innovation.

MARRIAGE AND THE FAMILY: ORIGINS AND FUNCTIONS

When asked why marriage exists, most laypersons will answer in terms of individual needs for love, companionship, sex, children, and homelife. Yet these answers are all time- and culture-bound to an extraordinary degree, and they ignore the fact that marriage is part of the social structure, not just a matter of individual option.

Every society we know of has some sort of prescribed type of marriage, whether it be monogamy, polygyny, or polyandry. Any member of that society must comply with the modal marital system and generally will live out his or her entire adult life within it. The marital unit is the primary economic unit in the society and the basis of all societal structure.

The argument that marriage exists so that society can regulate sexuality is an argument tinged with Victorian overtones of sexual repression, according to anthropologist Lucy Mair.[1] Most societies recognize some forms of premarital and extramarital sexual activities. In any case, defining sex as the focus for marriage is a relatively modern, Western idea, based on the equally culture-bound idea of romance and companionship as prime marital goals. As Ralph Linton put it:

> All societies recognize that there are occasional violent, emotional attachments between persons of opposite sex, but our present American culture is practically the only one which has attempted to capitalize on these, and make them the basis for marriage.[2]

One clue to the function of marriage in the social structure comes from the evidence of rules prescribing whom one is permitted to marry. In no society is one's choice of potential mates unlimited. All societies, for example, practice some form of incest taboo, which prohibits marriage (or sexual relations) between close kin (how close, and how kinship is defined, vary with the society). Scientific explanations for this tabloo range from Westermarck's idea of familiarity as the basis for sexual aversion to Freud's notion of incest as desired but discouraged by the Oedipal triangle and internalization. Nowadays, more societally focused views are gaining favor. Marriage is seen as a key part of the economic and political structure, cementing ties between families and groups. Lévi-Strauss, for example, has conceived of the incest taboo as originating in this need to develop a social network (and hence a society) through the ties of woman exchange. He quotes a tribesman:

What, you would like to marry your sister? Don't you want a brother-in-law? Don't you realize that if you marry another man's sister and another man marries your sister, you will have at least two brothers-in-law, while if you marry your own sister, you will have none? With whom would you hunt? With whom will you garden? . . . whom will you go to visit?[3]

The family of procreation, then, is seen by Lévi-Strauss as the basic unit of the social structure, while the socially reciprocal act of woman exchange is the definitive cultural bond holding societies together. In all societies it is the female who is exchanged between families or groups, not the male. We retain a vestige of this idea in our wedding ceremony, with the bride passing from the care of one male and lineage (her father) to another (her husband). Her name then changes from that of her father to that of her husband, thus symbolizing the exchange.

Anthropologists have often observed woman exchange, sometimes almost gleefully. The best-known case is that of the Tiwi of Australia, where wealth is defined in terms of control over women. The ethnographers Hart and Pilling note:

As in our culture, where the first million is the hardest to make, so in Tiwi the first bestowed wife was the hardest to get. If some shrewd father with a daughter to invest in a twenty year old decided to invest her in you, his judgment was likely to attract other fathers to make a similar investment.[4]

A female anthropologist observing the same people, however, looked on the system from a different perspective. Jane Goodale[5] emphasized the woman's power to define her own fate, and the tie of economic obligation between a woman and her son-in-law. It is possible, then, that anthropologists such as Lévi-Strauss have given us a biased view of the marriage system by emphasizing the woman-exchange aspect. One feminist critic, Gayle Rubin, has observed:

Kinship systems do not merely exchange women. They exchange sexual access, genealogical statuses, lineage names and ancestors, rights and *people*—men, women, and children—in concrete systems of social relationships. These relationships always include certain rights for men, others for women. "Exchange of women" is a shorthand for expressing that the social relations of a kinship system specify that men have certain rights in their female kin, and that women do not have the same rights either to themselves or to their male kin. In this sense, the exchange of women is a profound perception of a system in which women do not have full rights to themselves. The exchange of women becomes an obfuscation if it is seen as a cultural necessity, and when it is used as the single tool with which an analysis of a particular kinship system is approached.[6]

Susan Brownmiller, in her book on rape, *Against Our Will,* argues that men's control over women in marriage began because of the possibility of rape and the need for protecting women from it. Brownmiller comments:

Female fear of an open season of rape, and not a natural inclination toward monogamy, motherhood or love, was probably the single causation factor in the original subjugation of woman by man, the most important key to her historic dependence, her domestication by protective mating.[7]

She goes on to speculate:

The earliest form of permanent, protective conjugal relationship, the accommodation called mating that we now know as marriage, appears to have been institutionalized by the male's forcible abduction and rape of the female. . . . It seems eminently sensible to hypothesize that man's violent capture and rape of the female led first to the establishment of a rudimentary mate-protectorate and then sometime later to the full-blown male solidification of power, the patriarchy.[8]

Although Brownmiller's theory of the *origin* of marriage is not widely held by social scientists, little doubt exists about the importance of male regulation of sexual access to women in contemporary marriage systems. Many societies, especially southern European and Arab ones, define family honor in terms of the sexual purity of female kin. Severe sanctions accrue to transgressors of these codes. Such transgressions are often effectively prevented by the segregation or actual sequestering of women. The Islamic practice of *purdah,* for example, which mandates the veiling of women and their strict separation from public (i.e., male) life effectively removes sexual temptation by making extramarital liaisons almost impossible. Female genital mutilations, such as infibulation among the Sudanese, prevent premarital intercourse (and marital intercourse as well until certain remedial measures are taken).

Women and their sexulaity, then, are almost always subject to the authority of male kin, and the need to regulate sexual access to women often places major constraints on the life styles of women. Virginity is a female trait that is often valued (though not universally). On the other hand, male virginity is almost never a matter of social concern. Likewise, the typical culture is much more critical of extramarital sex among women than among men (especially when the former is revealed through illegitimate pregnancy).

In our society we think of marriage as being analytically distinct from childbearing or kinship networks. This view is clearly a modern luxury based on the possibility of birth control and the decreasing importance of the extended kinship network as the basis for economic and political ties. In most preliterate and traditional societies, the kinship network forms the basic latticework of the society, and individual identity is defined in terms of kinship ties. A man or woman is often not seen as fully adult until marriage or sometimes until childbirth. Anthropological theories of the origin of marriage, then, usually concentrate on the economic and child-rearing functions of the family unit.

One of the most widely cited views of the origin of the family is that of the anthropoligist Kathleen Gough.[9] According to Gough, the human infant's extended period of helplessness helped establish a sex-defined division of labor, with women concerned with child care and men with hunting and defense. Hunting over large territories was necessary since the distribution of game was sparse, necessitating long forays out into the bush. Such trips were impractical for women with small children and thus became the domain of the men. A survey of 175 societies by G. P. Murdock revealed that in 97 percent of them, men were the only hunters; in the rest, men were usually the hunters. Gough's idea is that "marriage and sexual restrictions are practical arrangements among hunters designed to serve economic and survival needs."[10] The parental pair with the ensuing kinship network forms the basis of social structure in such societies. The mother-child dyad may be the only constant across hunting societies, with about half opting for strictly nuclear families and the rest for some sort of polygamy. Every society recognizes some sort of social fatherhood, even when the scientific facts of the male role in reproduction are unknown. Gough remarks:

> Social fatherhood seems to come from the division and interdependence of male and female tasks, especially in relation to children, rather than directly from physiological fatherhood, although in most societies the social father of a child is usually presumed to be its physiological father as well.[11]

Even in hunting societies, however, women usually contribute a sizable share of the food supply through gathering or foraging for plant and vegetable matter that can be found in the bush. Such activities normally permit the simultaneous care and transport of young children, and in any case usually occur within a relatively small radius of camp, compared with the more widely ranging hunting parties. Such economically based theories of the formation of the family do not automatically explain the higher status of men and their power over women in the group. There is nothing "natural" about conferring higher status on hunting activities or those who engage in them, yet this almost always seems to be the case. Ernestine Friedl, an anthropologist, has provided an ingenious explanation of this process. According to Friedl:

> The most important differnce between the power of men and women to control others lies in the male monopoly of individual and small group hunts. This gives men the opportunity for large extra-domestic exchanges of meat, a source of power over others not available to women. The general principle that the generous distribution of scarce or irregularly available resources is a source of power, and that men and women differ with respect to their opportunities in this respect is . . . a significant element in understanding the roles of the sexes.[12]

Hunting also requires groups to work cooperatively to capture game, thus generating a network of social alliances not readily available to women. The gathering activities of the women and the centrifugal demands of child care serve to keep women apart rather than together. Male power seems most consolidated in groups that depend almost totally on hunting. Where plants gathered by women constitute an important part of the diet, their participation in the society seems more nearly equal. A study of the Kung bushmen[13] indicated that the sexes were more nearly equal in the nomadic bush settlements where women actively engaged in food gathering than in the sedentary horticulturally based villages where female participation in the economy was decidedly secondary.

In hunting societies, then, no matter how kinship is defined, power seems to be a male prerogative. Even when kinship is traced through the female line (matrilinearly) it is a woman's male kin who structure her life and choices. This power, according to the anthropological views discussed above, seems to derive from the male cooperative network of food distribution and prestige as opposed to the solitary female foraging activities. There does seem to be a pervasive distinction between the public domain of men and the private one of women that cuts across many different types of societies. However, to argue that men are more powerful because of their public life and women less powerful because of their private sphere borders on the tautological. Power is defined as being based in the public realm. Private power goes unrecognized, though it may exist.

The legendary "power" of women in patriarchal peasant and traditional societies comes to mind. Susan Rogers,[14] an anthropologist, has argued that it is women in such societies who actually exercise the greatest amount of power, but it is an informal sort exhibited in the domestic sphere and separate from the socially legitimate formal structure of village authority, which belongs to the men. She contends that anthropologists have only concerned themselves with the formal male authority structure and thereby underestimate the female presence. However, she admits that both male and female peasants believe in what she calls the "myth of male dominance," in which the male is acknowledged to be the head of the household and formal decision-maker. Other students of peasant societies disagree with this assessment, or at least think it is subject to considerable modification. Rayna Reiter, another anthropologist who has studied European peasant socieites, cautions against the overestimation of domestic power:

> The power associated with the private domain is limited when compared to that available in the public domain. A woman may reign within her family and have the power to control her children, but the state can turn them into soldiers or slaves. While women all over the world, and certainly in Colpied, [a French village studied by Reiter] derive informal prerogatives and status from their roles within the household, codified, legitimate power still rests in a realm from which they are usually excluded. To the extent that the family is a subordinate unit subject to the demands of the state, and to the extent that it is women who are defined by their role within the family, "separate spheres" can never be equal ones.[15]

We have slipped almost imperceptibly from a discussion of

the origins of the division of labor, sex roles, and public power in hunting and gathering societies to settled horticultural communities operating within the constraints of the modern nation-state. Perhaps we should pause a bit to understand the economic bases of sex role divisions in this very different economic system. Recall our earlier discussion of the role of child care in restricting female access to the public sphere of power by limiting hunting and associated public prestige rituals to men. Recall also that in such societies without modern birth control, women are likely to be almost continually pregnant, lactating, or caring for small children throughout their adult lives. What constraints, then, operated in settled horticultural communities to bring about the same division of labor and of power? According to Friedl,[16] men hold a monopoly over the clearing of land by cutting down trees and underbrush. Friedl attributes this monopoly to the fact that clearing land may require contact with potential enemies and warfare. She further argues that defensive functions more often fall to men than women because men are more expendable than women in terms of population maintenance since few men can impregnate many women and keep the population alive. We should also remember the discussion of aggression in Chapter 1, which suggested that there may be a hormonal basis for the male proclivity to aggression. In such horticultural societies, men also control the external trade system in which farm products are distributed, just as the male hunters did in hunting and gathering societies.

There is a basic problem, though, in assigning final responsibility to the economic system for the allocation of power according to sex roles. Most economic theories are ultimately based on what we can only call biological factors: for women, the constraints of child care and for men, the proclivity to aggression. If we also admit (as Friedl does not) that male dominance of hunting may be a likely consequence of greater male aggressiveness and physical size and strength, we have come almost full circle—back to a biological theory. It is biological, though, only to the extent that almost continuous female pregnancy and child care are biological "givens," since it seems that of the two, child care and aggression, it is the first that seems to have played the greater role in creating the sexual division of labor and the sex segregation of the public and private domains. The need to confer legitimacy on offspring by integrating them into a kinship network seems also to have contributed to the male public authority structure.

Societies such as our own, in which reliable birth control is widely practiced and small families are the norm, should presumably lack most of the constraints mentioned above. In addition, our society is one in which physical aggression and strength are not especially relevant to most public endeavors in the economic realm. Why then do such sex-linked divisions of labor and power persist? One possible answer is that they don't. And it is true that modern industrial society has seen unprecedented progress in women's rights and many incursions into formerly male spheres of power and influence on the part of women (although the male's economic support function has remained constant). However, one reason why this book was written is because of the need to explain the amazing stability of sex roles. Even in the last half of the twentieth century, in a world far removed from that of the hunters and gatherers and horticulturists, sex roles persist, and in fact the ideologies surrounding the roles (the masculine and feminine "mystiques") seem to have as strong a hold as ever on most people. To seek an explanation, we must make an excursion into the relatively recent history of the family.

Edward Shorter has written a persuasive and highly regarded account of the history of the family in his book, *The Making of the Modern Family.*[17] He outlines the development of the separation between home and work, which is basic to our modern conceptions of sex role spheres. According to Shorter, a profound "revolution in sentiment" in three key areas was responsible for our modern conception of the family. The first revolution in sentiment came in the area of partner selection when romantic love and sexual attraction became the bases for mate choice. Before the industrial revolution and the development of the capitalistic market economy, affection as the basic male-female marital tie was considered an aberration. According to Shorter, for most married couples, "it would never have occurred to them to ask if they were happy."[18] Some notion of individualism and freedom is a necessary prerequisite for this sort of bonding. Shorter's argument is that the "egoistical economic mentality" that underlies the market economy engendered "the sexual and emotional wish to be free." Young people entering the market economy were exposed to an individualism unknown to previous generations, and "in the domain of men-women relations, the wish to be free emerges as romantic love."[19] Once the bonds of the traditional agricultural community were loosened and young people migrated to the cities, the restraints of kinship and village life gave way to the ideology of the individual. Erich Fromm in his book *Escape From Freedom* has enumerated some of the negative, alienating consequences of this movement, which clearly laid the groundwork for the modern conception of selfhood and the consequent emphasis on individual happiness, and ultimately on romantic love.

The medieval idea of courtly love, of course, antedated the modern idea of love. Deriving from Arab roots, the elaborate codes of chivalry and the idealization of the female may bear superficial resemblance to modern notions of romantic love, but in fact they are quite different. Our conception of love as the basis for marriage presumes that sexual consummation and companionship are integral to its realization. Not so the courtly version. In *The Art of Courtly Love* it is stated quite clearly:

> . . . everybody knows that love can have no place between husband and wife. They may be bound to each other by a great and immoderate affection, but their feeling cannot take the place of love, because it cannot fit under the true definition of love. For what is love but an inordinate desire to receive passionately a furtive and hidden embrace? But what embrace between husband and wife can be furtive, I ask you, since they may be said to belong to each other and may satisfy all of each other's desires without fear that anybody will object? Besides, that most excellent doctrine of princes shows that nobody can make furtive use of what belongs to him.[20]

The adoration of women from afar that the code of courtly love entailed can perhaps best be understood as being more closely akin to the cult of the Virgin Mary (which flourished at the same time in Europe) and to the later Victorian reverence for women on a pedestal (as in the cult of Southern womanhood) than to our modern conceptions of romantic love as a basis for marriage. Critical to the modern conception is the liberation of feelings (especially sexual ones) and the idea that men and women can be companions to each other in their mutual search for individual happiness. Such notions of individual happiness and sexual friendship could only flourish in a post-traditional society where the claims of the community had taken a decided back seat to those of the "free" individual. Women as well as men were entering the capitalistic marketplace and were subject to such changes. According to Shorter, this first revolution in sentiment, toward romantic love as the basis for marital choice, originated among the proletariat and filtered upward.

The reliance on romantic love for marital choice did not mean that it was the sole criterion. Indeed, economic and other more practical factors entered in and still do to a large degree. But within the broad constraints of class suitability, a new factor, that of love and companionship, entered to form the psychological basis of marriage. Ironically, according to Shorter, it was probably this factor of romantic sentiment that began the rush toward domesticity and the ultimate segregation of women from public life, even in industrial society. For romantic love laid the groundwork for the home as an emotional retreat and led to the other two revolutions in sentiment: maternal love and domesticity, which came to be major defining factors in the sex role of modern women. Before the home became sanctified as a retreat, women and men had worked in separate domains, to be sure, and women's work was more closely confined to the household and the care of children. But the three revolutions of sentiment and the subsequent development of modern family life led to the psychological retreat of the woman into the home and to her definition in terms of its expressive functions even when she is not actively engaged in homemaking and child care. Thus, the "feminine mystique" Betty Friedan wrote about is an ideology of the happy home as a retreat from the marketplace, with the female as the keeper of the hearth. This definition of the female is a product of the industrial age and the modern mentality and definition of the family.

The second great revolution in sentiment, according to Shorter, was the injection of sentiment into the mother-child relationship. "Good mothering is an invention of modernization,"[21] Shorter contends. Of course, we tend to think of mother love as instinctual, or at least as a normal component of every socialized female's relation to her offspring. Yet, Shorter argues that this sentiment, like romantic love, is a product of an age; a historical consequence, not a natural display. Many arguments, of course, for essential female character traits (i.e., Helene Deutsch's) give a central role to maternal love; now we have its cultural relativity suggested. Shorter gives evidence of widespread neglect and abandonment of children in premodern Europe. The general use of hired wet nurses (who often neglected and abused their charges) underlined the lack of bonding between mother and child, or indeed among any family members. Other historians of the family have argued that high infant mortality led to an absence of sentiment (one could not invest emotionally in a child who was unlikely to survive). Yet, Shorter convincingly argues that:

> The high rate of infant loss is not a sufficient explanation for the traditional lack of maternal love *because precisely this lack of care was responsible for the high mortality.* At least in part, if children perished in great numbers, it wasn't owing to the intervention of some *deus ex machina* beyond the parents' con-

trol. It came about as a result of circumstances over which the parents had considerable influence: infant diet, age at weaning, cleanliness of bed linen, and the general hygienic circumstances that surround the child—to say nothing of less tangible factors in mothering, such as picking up the infant, talking and singing to it, giving it the feeling of being loved in a secure little universe. Now by the late eighteenth century, parents knew, at least in a sort of abstract way, that letting newborn children stew in in their own excrement or feeding them pap from the second month onwards were harmful practices. For the network of medical personnel in Europe had by this time extended sufficiently to put interested mothers within earshot of sensible advice. The point is that these mothers did not *care,* and that is why their children vanished in the ghastly slaughter of the innocents that was traditional child-rearing. Custom and tradition and the frozen emotionality of ancien-régime life gripped with deadly force. When the surge of sentiment shattered this grip, infant mortality plunged and maternal tenderness became part of the world we know so well.[22]

Shorter's view, then, is that sentiment preceded infant survival rather than the other way around. The modern family is of course unthinkable without the core of husband-wife and parent-child bonding. Doubtless, the preceding revolution in romantic love lay the groundwork for the second one, in maternal affection, since one would expect feelings to spill over from the sexual bond to the product of that sexual union. Even when the marital bond was in reality not aglow with romantic ardor, the society's new normative expectations about family links probably served to generate similar expectations about parent-child ties. The limitation in family size brought about by birth control probably also served to change the image of the child from that of an unwanted by-product of sexuality between two not especially well-matched adults to a loved center for the family unit.

The foundation laid by romantic and maternal love made probable the transition to domesticity, the third revolution in sentiment, and the definition of the nuclear family as we know it today. According to Shorter, the nuclear family is "a state of mind rather than a particular kind of structure or set of household arrangements."[23] It is characterized by "a privileged emotional climate" and "a special sense of solidarity that separates the domestic unit from the surrounding community."[24] Preindustrial family life was much less private and not at all cut off from community life. Missing was the sense of emotional retreat we so closely identify with modern family life. The home itself became more private, with separation of rooms and a new emphasis on privacy. The third revolution in sentiment, then, according to Shorter, revolved around this burgeoning of domesticity.

Of what consequence for sex roles is this turn in the history of the family? Although it seems a change of great significance, perhaps it is somewhat less important than might at first appear. Throughout changes in family style, the subordinate position of women in Western society has remained constant. When the primary site of labor was the home and surrounding fields (as was the case in preindustrial society), women were part of the labor force, although some allowance was usually made for physical strength differences and child rearing. Women were not actually dependent on their husbands' labor, but they were still largely defined by their family membership and were definitely considered to be second-class citizens in the public world. The medieval guild system, for example, was all but closed to women, and the legitimizing influence of Catholicism and Protestantism in maintaining sex role divisions must be acknowledged as a factor.

Women could hire themselves out to other families as domestics and could expect limited entry to some trades. By and large, though, they were not seen as independent entities, but rather as dependents of men, despite their economic participation. The change that the rise of the sentiment-based nuclear family brought about was to delineate the economic dependence of women on men (i.e., their husbands) much more sharply and to foster a societal ideology of domesticity and family life that led to the "ideal" of the leisure-class housewife tending the needs of her family. Despite increasing female participation in the labor force throughout the twentieth century, which in reality belies this definition of women, the ideology still exerts a tremendous psychological hold over men and women and affects the achievement goals, life plans, and actual occupational choices of women in far-reaching ways. Likewise, the lives of men are shaped by this family vision, since it is they who are the externally anchored economic support for the warm little family nest.

The modern family, then, changed from an economically defined unit comparatively bereft of any emotional ties to a child-centered emotional retreat from the intrusions of the community. This retreat was to be presided over by the wife and mother and to be economically supported by the market labor of the husband. The favorite description (originated by Parsons and Bales) of the sociological interplay between the female expressive role of emotional support and the male instrumental role of a task-oriented nature that is supposed to make the family "work" is based on the supposition of domesticity as the core of family life.

The withdrawal of the family from the community and public life presaged the psychological seclusion of women from such life. Although women have always been somewhat restricted from community participation, the new notions of family life provided a new rationale for their fate. It made the subordinate female role into a desired cultural goal for women. Even when participating in the public domain, women were largely defined by their role in the private one.

The main lines of argument outlined above were developed by Frederick Engels in his well-known work, *The Origin of the Family, Private Property and the State*. Engels championed a much-discredited historical progression from group marriage to modern monogamy, singling out the advent of private property and the capitalistic state as the cause of the oppressed state of women. Drawing from the evidence of kinship terminology and the work of the American anthropologist Lewis H. Morgan, Engels concluded that a stage of group marriage ensued after an initial state of sexual promiscuity. During the stage of group marriage, females possessed a great deal of power, creating a matriarchal family system. Women at that time were primarily employed in the household, but they derived a great deal of power from their maternal position in the structure of the extended household. Since the society as a whole existed on a subsistence level, male productive skills did not count for much. With the creation of a surplus by improvements in agriculture, the male became more powerful and established a patriarchal family system to pass on private property to his heirs. For this to be possible, the family had to be defined in terms of the male line. The family became privatized and removed from the communal household organization of old. Engels writes:

> In the old communistic household, which comprised many couples and their children, the task entrusted to the women of managing the household was as much a public and, socially necessary industry as the procuring of food by the men. With the patriarchal family, a change came. Household management lost its public character. It no longer concerned society. It became a *private service;* the wife became the head servant, excluded from all participation in social production. Not until the coming of modern large-scale industry was the road to social production opened to her again—and then only to the proletarian wife. But it was opened in such a manner that, if she carries out her

duties in the private service of her family, she remains excluded from public production and unable to earn; and if she wants to take part in public production and earn independently, she cannot carry out family duties. And the wife's position in the family is the position of women in all branches of business, right up to medicine and law. The modern individual family is founded on the open or concealed domestic slavery of the wife, and modern society is a mass composed of the individual families as its molecules.[25]

The historical and anthropological basis of Engels's developmental theory seems to be quite shaky; however, his characterization of the modern family accords quite well with that of other historians of the family. Engels used kinship terminology as evidence for communal sex arrangements. For example, some common terms were applied to all same-sex kin of the parental generation, (i.e., one term stood for all the father's brothers and his father and to all mother's brothers). Modern anthropologists disagree with Engels' conclusions from this evidence, and also doubt that any society was ever truly matriarchal. As we have seen in an earlier part of the chapter, primitive hunting and gathering societies operate on the basis of male public power and authority. Engels' use of the analysis of myth and ritual by such writers as Bachofen and Briffault to argue for an initial stage of matriarchy has not been upheld by contemporary anthropologists. The presence of fertility cults, for example, is not deemed to be relevant to the power of women in the society.

Engels argued that before the advent of private property, women's household work was valued as much as men's productive work and hence conferred equal (or even greater) power to women. As we have seen, anthropological evidence from primitive societies does not support this view, since a pervasive male power structure seems to be the rule. However, in societies where women contribute a larger portion of the food supply, their power seems to be greater.

We do find corroboration for the latter part of Engels' theory, which does seem to coincide with Shorter's analysis and that of other historians of the family. The market economy and the advent of capitalism seem to have generated a succession of family changes culminating in the idealization of the nuclear family's domestic seclusion and the division of the sexes into female expressive and male instrumental roles within that family system. Even when the female participates in the labor force, the normative view of her (and very likely her view of herself as well) is a domestic one. It is perhaps this cultural and psychological image of womanhood and consequent socializing practices that go the furthest in explaining the nature of the current sex role system. The male, in a complementary fashion, defines himself almost exclusively in terms of his role as economic provider for his family and is likewise socialized to see the sex role world in this way. The new ideology of domesticity, then, did not create the sex role systems, but rather built upon already existing cultural ideals of the sexes and added a new element to their maintenance. The result is that even when women participate in the public sphere (a relatively new development) they are defined by their role in the private one.

Let us return from our excursion into history and anthropology and again consider the main problem under discussion: the nature of the current sex role system and its maintenance processes. To do this, we will discuss the division of labor according to sex roles (domestic and occupational), and finally, the socialization processes that maintain this system.

THE DIVISION OF LABOR ACCORDING TO SEX ROLE: DOMESTIC AND OCCUPATIONAL

The American ideal was to catch a man
before you were too old, say twenty-two,
and to take a deep breath,
disappear into a suburban ranch house
and not come up for air until your children
("a boy for you a girl for me")
were safely married.

—Maggie Tripp
Woman in the Year 2000[26]

The sentiments voiced above represent a very real statement of this culture's view of the division of labor according to sex role. (The silent part of the above vision of course is the successful businessman dashing out the door to catch the 8:18 commuter train, on his way to spend the day in economic pursuits, secure in the knowledge of his domestic nest as a relief from such pressures.) The statistical reality of such a vision is secondary. America is for the most part a society that furthers a middle-class view of life featuring a healthy (or unhealthy) sprinkling of material comforts and psychological security. This vision is promulgated over and over again through the media (especially television) and directly in the socialization of children in the home and school. For the most part, social mobility in this society means participation in this "American dream." Such a vision has a tremendous psychological hold on individuals and is a definite factor in the shaping of life plans. Such life plans are sharply demarcated by sex, and form the core of the American sex role system of the division of domestic and occupational labor.

As we have seen in our survey of the history of the family, the ideal of domesticity is relatively new. Accompanying it has been the definition of women as ideally constituting a leisure class devoting their time to the management of the home and care of children. Of course, these two pursuits have always been a part of the female sex role, but as we have seen, they were not commonly viewed as the defining qualities of female life until relatively recently. Heretofore, economic activities in and around the home, in agriculture, and in the early cottage industries, were assigned to women. In such societies, women still played a decidedly secondary role and were confined to the private sphere. With the coming of the Industrial Revolution and the separation of the workplace from the home—and with the new definitions of domesticity—women were gradually weaned from the economic work force. Recently, there has been a decided return to the workplace by married women (who with the help of birth control are not tied to the lifelong cycle of pregnancies and child care). However, women who have returned to the workplace have returned to secondary positions, and the residual effect of the domestic value complex is obvious: female workers tend to view themselves and to be viewed by others as being primarily defined by their family rather than by their occupational role. (The case for men is, of course, the reverse.)

The census statistics on working women tell an interesting tale.[27] In 1800, women made up 4.6 percent of the work force; in 1900, 13.3 percent; in 1950, 27.9 percent; in 1960, 37.1 percent; and in 1976, 40.7 percent, a figure that was originally predicted for 1985. Studies of cohort groups, that is, those that follow one group of women born in a given year throughout their life cycle, show that at virtually every age, women born more recently are more likely to work. At age thirty, for example, only about a fifth of all women born between 1886 and 1895 were working (during the period 1916 to 1925), while nearly half of all women born between 1936 and 1945 were in the labor force at age thirty (which was, in real time, the years between 1966 and 1975). In 1973, 44 percent of American women worked either full- or part-time, and by mid-1976 this figure stood at 48 percent. The most substantial gain over the years in female employment has been in the labor force participation of married women. Single, divorced, and widowed women have always figured in the work force because of their need to support themselves; the appearance of married women marks a decided break with tradition. The end of the sex role system, right? Women in the labor force, men in the labor force, no home-work split, right? Wrong, read on.

Lurking behind the statistics is another story, one of sex role division and tradition. Three areas are notable here: (1) the life- (or, better, family-) cycle dependence of women's work as opposed to men's; (2) the considerable income differential between male and female workers, reflecting the differences in occupational status and prestige; and (3) the preponderance of part-time, as opposed to full-time, work for women. All of these factors relate to the differing life plans of men and women with regard to the occupational and domestic spheres, a major component of the current sex role system.

The most glaring sex role difference, which in a sense is the precursor of all the rest, is the first: the broken nature of female work force participation. For example, if we look at married women (husbands present) with children under three, only 15.3 percent of them were working in 1960. By 1974, 31.0 percent were in the labor force. However, being "in the labor force" does not mean holding down a full-time job. If we look at 1959 statistics, we find that only 26.6 percent of married women without children were working full-time, while for women with one child under the age of two, the rate was only 6.8 percent.

A study by Garfinkle, based on 1960 census data, indicated that:

> The birth of a child reduces the average number of years a married woman can be expected to spend in the work force by about ten years. The birth of each additional child appears to further reduce the work life expectancy from two to three years for each child.[28]

Here, then, the practical issue of "who shall care for the children?"[29] intrudes. When we look at mothers who do work, we find an interesting pattern of child care alternatives. In 1970, of mothers with children under six, about half were cared for in the home, usually by the father or another relative. About ten percent were cared for in a day care center and 35 percent in someone else's home, either by a relative or by someone employed in a private informal child care arrangement. In the last five years, the number of mothers using day care centers has almost doubled; likewise the proliferation of care outside the home. The latter development was probably helped along by the decrease in the number of homes having extended kinship members outside the nuclear family. When children reach school age, many mothers solve the child care problem by working only during school hours (43 percent did so in 1970). Of course, the number and type of jobs that fit that criterion are severely limited, as are the career (as opposed to job) options such a choice affords.

The effect of children on female work styles goes far beyond the limits of the actual requirements of child care. Although not an insignificant problem, the actual demands of child care could be met by a network of child care facilities; and in any case the problem is self-limiting in any given case since the number of years needed to care for a small child does not constitute a large part of a thirty- to forty-year work life. The real problem lies in the effect socialization for the maternal and domestic role has on shaping the life plans of women with regard to career choice. The cultural values of domesticity and maternity inculcate the vision of a woman's life devoted to such pursuits. For a large proportion of women, especially in the critical early years of adolescence and young adulthood, these goals seem to overshadow all others and preclude training and orientation toward a career goal. After marriage and child rearing, when the woman is ready to enter (or reenter) the work force, she often finds that her training is inadequate for all but the lowest rungs on the job ladder. The motivation of men to enter career fields comes about because of the considerable cultural "push" toward male achievement, a pressure missing in female socialization.

There is a very real concern on the part of many women and men as to the adequacy of any but maternal care for young children. Acceptance of this idea would tend to support the primacy many women give to this goal, which results in the relative neglect of their own career ambitions. Recent studies of the effects on the child of a working mother have not come up with the predicted bad effects. According to a recent review by Hoffman and Nye:

> The data, on the whole, suggest that the working mother who obtains personal satisfaction from employment, does not have excessive guilt, and has adequate household arrangements is likely to perform as well as the nonworking mother or better; and the mother's emotional state is an important mediating variable.[30]

The skeptical reader will note the many "ifs" in the above statement. The mother's attitude, however, is likely to be a primary factor, both for the stay-at-home mother as well as for the working one. A mother who is bored and unhappy with housework and child care is apt to transmit negative feelings to her children and might be better off working. The difficulties of child care should not be underestimated, however. But two points need to be made to clarify the issue. First, why is child care defined as a female responsibility, and why does it interfere with female career development, not male? Second, as we stated earlier, the "problem" of child care is self-limiting in any given case, and with today's smaller families should only take up a small part of the woman's actual life span and potential work life. As it is, though, such pursuits seem to define female life goals, although there is a growing trend away from such an ideology.

It would probably take a very radical societal shift in sex role conceptions to define child care as appropriate for any persons other than females. In all known societies, women bear the primary responsibility for child care. Therefore, for the issue of female employment, the facts of childbearing are likely to have a very heavy input for some time to come. In some sense, though, women are being socialized for a somewhat obsolescent role. Over the past century there has been a steady decline in the birth rate; the average number of children now hovers around two. And with increasingly reliable birth control techniques, the desired and actual family size are likely to coincide more closely than ever. Concern over population pressures is likely to be one factor, but perhaps a more important one is what John Scanzoni has termed the value complex of "individualism" triumphing over that of "familialism." He has evolved a model of a female attitudinal and behavioral complex that promises to presage considerable changes in the nature of the family and of female participation in the work force. According to Scanzoni:

> The model begins with full-time premarital employment, which appears to be influenced positively by education. Employment, education, and sex role modernity combine to increase age at marriage. Each of these factors tends to increase the likelihood that wives will work full time after marriage and prior to children. In turn these cumulated factors tend to increase the spacing interval between marriage and the first child. Next, all the preceding reduce the number of children born to these younger wives. Sex role norms, greater education, employment behaviors, deferral of marriage and of the first child—all reflect preferences for individualistic gratifications. The stronger these preferences are, the fewer children or familistic gratifications these wives have. In contrast, the weaker these preferences the greater their level of familialism. One set of rewards functions as compensations or alternatives to the other.[31]

The age of marriage, especially for females, has increased, and so has the tendency of young women to live on their own before marriage. Such experiences and opportunities for independent living and work experiences will undoubtedly have an impact on their life plans with regard to career and marriage.

For a long time, the correlation between female fertility and participation in the labor force has been noted. That is, women who work have fewer children than those that don't, or, to put it another way, women who have fewer children are more likely to work than those with more. The direction of causality in such a relationship has been in dispute, however. Several possible interpretations come to mind. Women could be limiting their families to be compatible with their work expectations, or women could be shaping their work lives around their fertility. Which factor is primary, plans for future labor force participation or fertility expectations? Scanzoni's model, considered above, discusses the two factors concurrently and does not seem to take a position on the primacy of one over the other. By using a nonrecursive statistical model that they believe will permit them to assign a causal direction to the relationship, Waite and Stolzerberg[32] have concluded that it is the woman's plans for work force participation that shape her fertility plans rather than the other way around. Such an argument would be compatible with the idea that increased education and early labor force participation alter the balance in favor of career rather than domestic values, and the fertility plans come along as a natural consequence of such an orientation.

Presumably, an increased orientation toward working in general will bring along with it an increase in the choice of careers at higher rungs in the occupational hierarchy. Such a shift in female choices is not just critical in individual lives, but in the sex role system in general, since it is through occupational achievement that societal status and prestige are attained. As long as women remain in the home and outside the labor pool, they will forever be left out of this societal ranking system, gaining any status they have solely through the positions of their husbands.

The money gained through female employment is of course not an insignificant factor in the sex role system. When asked their reasons for working, most women list financial concerns first. The economic independence that such employment brings (even though it is usually less than that of the husband) may change the balance of power in the home away from the male-dominated model to a more egalitarian arrangement. Many feminist thinkers have stressed the psychological consequences of such employment, from Charlotte Perkins Gilman to Betty Friedan. Gilman, writing in 1898, said:

> Not woman, but the condition of woman, has always been a doorway of evil. The sexual-economic relation has debarred her from the social activities in which, and in which alone are

developed the social virtues. . . . In keeping her on this primitive basis of economic life, we have kept half humanity tied to the starting post, while the other half ran.[33]

Social trends, then, toward increased education for women, more work experience, delayed marriage, and decreased fertility seem to be pointing the way toward greater labor force participation throughout the life cycle. Another factor that should not be neglected is the role of labor-saving devices for domestic tasks (we are going on the traditional assumption of assigning such tasks to women since in the present sex role system, this is, by and large, the way things are). The introduction of ready-made clothing, of washing machines, dishwashers, vacuum cleaners, and convenience foods have all contributed to the reduction in women's actual workload. However, the important role of attitudinal factors must be mentioned. A survey by Joann Vanek[34] revealed that employed women spend an average of twenty-six hours a week on housework, but that nonemployed women spend fifty-five hours. The latter figure is equivalent to that reported as average for housework fifty years ago! An important part of the female domestic value system (what Betty Friedan dubbed the "feminine mystique") concerns itself with the perfectionist management of the household. Women who subscribe to such a constellation of values are likely to find enough to do at home to round out a full week's work and more. Working women, on the other hand, are likely to be more efficient in the use of their time and to be able to pare down the work to a minimum.

Work participation alone will not change the nature of female status or of the sex role system. Women do work today in greater numbers than ever before, but they tend to work in secondary positions outside of the societal ranking hierarchy. Secretarial work, for example, is an almost totally female monopoly, while business management remains a male preserve. In 1974, the average female wage ($6,957) was only 57 percent of the average male salary ($12,152).[35] To understand the occupational socialization system, we shall briefly review two concepts that have been actively advanced in the literature as psychological explanations for lower female occupational achievement, and then substitute our own, that of differential life plans based on the (perceived) nature of the family system.

The two concepts that have received the most publicity in psychological circles are: (1) achievement motivation, and (2) fear of success. The first concept, developed by David McClelland[36] and his colleagues in the early 1950s, refers to a person's need to compete against a standard of excellence. The conventional way of measuring such motivation is to give the person a series of pictures (called the Thematic Apperception Test, or TAT) about which stories are to be composed (one to each picture). A typical picture would have a boy playing a violin gazing out a window. The stories are then scored for "achievement imagery" to see if the person seems to be concerned with achievement and competition. For example, if a person said "The boy is dreaming of becoming a great violinist," that response would score high in such imagery, and so on. Men consistently outscore women on such tests. It must be noted, however, that a good proportion of the images concern men in potential achievement situations rather than women, and this may bias the results. Let us accept for the moment, though, the truth of the proposition that men in fact seem to see more achievement themes than women. This result should not surprise us, given the typical sex-typed socialization experience boys and girls receive.

We may want to call the result of this socialization a motivational system, but it makes just as much sense to think of it as a different life plan. In the case of males, this life plan is broadly economic and competitive; in the case of females, domestic and expressive. Early independence training seems to correlate with high achievement motivation; therefore changes in female socialization toward such values would very likely have repercussions in female achievement as well. We shall consider in more detail later the career-choice process for women and shall see that the concept of differential life plan is probably a more convenient rubric than achievement motivation. Men achieve not because of high achievement motivation but because they *have* to; that is their life plan in relation to the family system. They then score higher on achievement motivation tests because they have learned to interpret the world in achievement-relevant terms. They have learned to do this because that is what their life is all about, and that is what they were taught as children.

Another set of findings about achievement has been used to explain lower female career attainment.[37] An attribution study found that women tend to explain their achievements (e.g., doing well on a test) in terms of luck rather than skill, while men use skill as an explanation rather than luck. Such external attributions for women might tend to foster a more negative self-image with regard to their own skills and self-confidence in using them. A further set of findings seemed to support the notion that women are more disturbed by failure in achievement-relevant tasks than men are, hence they are less persistent in the face of discouragement (a necessary concomitant of any career climb). Such studies, however, seem to show a turning away from achievement orientation, or what I would prefer to call occupational life plan, rather than explaining its cause.

A second motivational state posited in the literature has been that of fear of success, originated by Matina Horner[38] (now president of Radcliffe College). According to Horner, achievement situations are conflict-laden for women and arouse a competing motive to that of achievement. This competing motive is the avoidance or fear of success. It arises out of the negative image of successful women inculcated by our societal value system. Horner originally gave a short sentence to her subjects about a woman being at the top of her medical school class and scored the subsequent stories made up in terms of fear-of-success imagery. According to Horner, the replies fell into three major categories: fear of social rejection, concern about one's normality or femininity, and denial of the achievement situation itself. An example of the second category should suffice to give the flavor of the concept:

> Anne is completely ecstatic but at the same time feels guilty. She wishes that she could stop studying so hard, but parental and personal pressures drive her. She will finally have a nervous breakdown and quit med school and marry a successful young doctor.[39]

Recent studies of this concept have failed to substantiate its independent existence or its greater occurrence among women. However, anyone with any experience among young women can impressionistically substantiate the nature of the concept, which, of course, relates to the life plan of domesticity rather than occupational achievement. Many of the factors that we shall examine in our model of career development concern the muting of these concerns over such an "unfeminine" choice. Gisela Konopka has studied adolescent girls from a variety of backgrounds and found that many struggled with the concept of combining marriage and career. In the absence of social supports, it is often the latter that is compromised. For example, one of Konopka's fifteen-year-old suburban respondents spoke of her career plans:

> Maybe a secretary or something like that. Hopefully I will be married with some children. That's what I want to do, I guess that's the way I've been brought up. I don't want to have any

career. We started having career reports at school and I was at a loss of what to do because I don't want a career. I do want to have something that I can do and I do want to get married and have children. I do want to have something to fall back on if something happens, or if I decide to go back to work.[40]

It was this young woman's early socialization to a domestic life plan that led her to screen out new, conflicting sources of information. The post hoc quality of her career planning will ensure her a place on the lower levels of the career hierarchy. There is a decided trend for women now to plan for more education and career attainment, but the pull of the domestic or nonoccupational life plan as primary is considerable. We will later consider the nature of socialization for the occupational life plan for men, but will first turn to the nature of the domestic plan for women. This life plan becomes such a goal in the adolescent and early adult years that it prevents occupational preparation and participation for a large proportion of women. Such participation, as we have stated earlier, is critical to changing the sex role system and the position of women, since occupational entry is prerequisite to status and power in this society. Any large change in women's entry to careers will have an impact on men's socialization and life plans; but we will postpone discussion of this until after our consideration of the nature of the female life plan of domestic rather than occupational participation.

FEMALE CAREER CHOICE

Perhaps the best place to start in trying to characterize the sex-linked nature of the process of career choice is in an examination of the meaning of *choice* itself. *For women, choice of any particular career is confounded with the choice of having a career at all.* Thus, as Lotte Bailyn states, "in making decisions about a style of life, a woman must choose in ways that men do not choose. . . . For most men there is no basic choice as to whether or not to work. . . . But for a woman, society creates not a decision but the necessity for a choice. She must decide whether to include work in her plans and if so, how much of her life she should devote to it."[41] Because the decision to have a career at all in some ways overshadows the commitment to any particular career, we shall focus on those considerations common to all career choices and discuss differences between decisions to enter particular careers only when necessary.

In a short study I once did of the sequence of career choices among women graduate and law students, I found that very early career choices (up to ages ten to twelve) tended to cluster around traditional female roles (e.g., teacher, nurse) and then go through *several* professional (nontraditional) careers in the mid-teens (e.g., physician, college teacher) before settling on the final one, which may suggest that the decision to have a professional career at all is logically prior to that of entering any particular career. (Some women, of course, may "hit" on the right career the first time and stick to it, but the pattern of several professional career choices seems to be the mode.) Thus a protocol from a first-year law student reads: nursing (age ten); teaching in high school (twelve to fourteen); teaching at the college level (fifteen to seventeen); lawyer (eighteen to nineteen). A second-year law student's sequence is: policewoman (eight to twelve); physician (twelve to seventeen); nuclear physicist (seventeen to eighteen-and-a-half); economist (eighteen); lawyer (nineteen-and-a-half). (The continuity in interest for this respondent between policewoman and lawyer is also of interest, though this may reflect selective retention of now relevant very early career choices over irrelevant ones).

What factors determine which girls make the "switch" in their teens? It can be explained in part by the broader world of the high school curriculum, which introduces new areas (e.g., the sciences) to the student; but why do some girls see these areas as relevant to their future careers while others do not? I would propose that such "readiness" is determined by both sociological and psychological variables, but that it is very much influenced overall by the prevailing sex role standards in our society. It is to this model that we now turn.

The values subsumed under the general rubric of the "feminine mystique" (a term coined by Betty Friedan in her 1963 book of the same name) constitute one major cluster of relevant factors, and it is differential transmission of these values that is a prime determinant of the decision to pursue a career or not. The "feminine mystique" is a loaded term that is shorthand notation for the American female sex role standard as it concerns domestic values. Its prime components include the idea that the adult roles of each sex are very different, with the woman's primary (and perhaps only) role being in the home, with the corollary that she is expected to find fulfillment through the lives of her husband and children rather than in a career. Such a value complex clearly specifies different life plans for women and men. The mystique is supported by different educational goals for women (oriented toward marriage) and by systematic discouragement of deviance (indirectly the negative stereotype of the striving career woman; more directly, job discrimination—or perhaps more accurately, *career* discrimination), and the mass media (the glorification of housework in commercials and the absence of women in other than traditional roles).

Though there is little disagreement that these values exist in the culture and that their logical implication is that a choice against a career should be made, clearly there is differential transmission of these values to girls in their formative years. Many of the sociological correlates of career choice can be seen as variables affecting the adoption (or perhaps internalization) of the mystique. We shall try to understand the following factors in this light:

1. *Role models of succesful career women* (especially maternal role models). This variable comes up time after time in (the few) studies of career choice in women. We can understand the influence of role models in two ways: (1) If the role model is the mother, she is less likely to transmit the values of the mystique to her daughter since it would be a contradiction of her own way of life; (2) Whether the role model is the mother or someone else, the girl is faced with a concrete contradiction of the values of the mystique.

Maternal role models (even if they are just *working* mothers rather than *career* mothers) seem to be particularly important.

Thus Carol Lopate[42] reports that among women who dropped out of medical school (mostly for nonacademic reasons), two-thirds had mothers who had not worked, whereas among those women students who continued their studies, only half had nonworking mothers. This finding implies that those women with working mothers were better able to handle the stresses of entering a career in a traditionally masculine field. A similar finding was reported by Herzfeld.[43] She administered a self-concept test to a sample of Radcliffe students and found that those with working mothers maintained a more favorable self-concept and seemed to be undergoing less emotional strain that those whose mothers had not worked (with social class, religion, and education controlled). She explains this finding by pointing to the pervasive career orientation among Radcliffe students and suggests that this atmosphere puts a strain on students without suitable maternal role models. Interestingly enough, it did not seem to matter *when* the mothers had worked (most had worked before their daughter's birth or in her childhood), indicating that differential transmission of the mystique per se rather than the present role model might be the important variable.

Ginzberg[44] found that the mothers of 75 percent of his sample of Columbia women graduate students had worked at some time. This percentage is higher than the average for that social class in the first few decades of this century (when the mothers would have worked). Lopate, in her report on medical students, found that the average maternal educational level is higher (some college) for women medical students than for their male counterparts (high school graduate). Although this educational factor does not guarantee a higher percentage of working mothers, the trend is for a greater proportion of more educated women to work relative to the less educated. Phoebe Williams,[45] in a study of Radcliffe women (in graduating classes from 1906 to 1966) who went on to medical school, found that only 46 women in her sample of 203 had mothers who had never worked. Eighty-four of the 203 had mothers who had worked at some time after they were born, and the rest (72 plus one no-reply) had mothers who had worked only before they were born. The mothers who had never worked were concentrated in the very earliest group of alumnae (early 1900s), so that for the later groups, the percentage of working mothers was much higher. Information on the employment status of mothers of comparable Radcliffe students who did not enter careers was not available, but it is likely to have been lower.

Alice Rossi's[46] study of women college graduates of the class of 1961—which she divided into marriage and career "types"—does not indicate the family background of her sample, but she notes that career types disliked such family activities as visiting with relatives more than marriage types, and one might infer that their mothers also had not been exclusively family-oriented, perhaps as a result of work commitments. Rossi adds:

> Women whose childhood was characterized by intimate and extensive relationships with their familes, with relatives as well as members of their immediate nuclear family, are far more apt to grow up with a very conservative image of appropriate roles for women.[47]

One should not neglect the important point that mothers in traditional roles can act as *negative* role models if they are notably unhappy in their limited role and lack of career. Ginzberg quotes one woman: "My mother's lack of an occupation convinced me in the sense that it made me want one."[48] The mystique can be negated as much by a negative example as by a positive example of a counter-case (working mother); but one suspects that a positive role model (e.g., a teacher) might have come in at some point.

Role models outside of the family may also be important influences. Thus women college professors may influence the career decisions of their women students (although the groundwork for receptivity to such influence was probably laid earlier). However, this process may backfire, since many female college professors of middle age and beyond never married and thus may present *negative* role models to aspiring female professors. Although some students may still be recruited on the old model (what Bernard[49] calls "the vocation for celibacy"), most would be discouraged by such a model. Older, unmarried female faculty members really present an inaccurate role model, since a much higher percentage of younger academic women marry.

Role models probably enter into women's career choice unconsciously more often than consciously and may provide the necessary "push" to continue pursuing a career when the temptation to follow the mystique comes (e.g., see the Lopate discussion of medical school dropouts mentioned earlier). Sometimes the use of the role model may be more conscious, however. Lopate, for example, reports the emulation of successful women doctors as a common motive among women medical students (occasionally the role model is also the student's mother). A law student in my survey traced her decision to: "an inexplicable interest in a "notorious" criminal case in 1960 (Caryl Chessman)—which I believe was a miscarriage of justice—led to my determination to fight such evils through a career in criminal law. Also the fact that there was a woman lawyer in that case was important."

Role models may be useful as decision reinforcers at later points, especially immediately prior to and following career choice. Therefore the presence of female faculty members and other training models is an important ingredient in successful female career socialization.

2. *Paternal attitude* (and, by extension, that of other men). This factor is complementary to maternal influences, as discussed above, for two reasons: (1) it can directly affect career choice through nontraditional socialization practices and through exposure to career fields, perhaps the father's own; and (2) it indirectly provides a counter-model for the cultural "male attitude," reassuring the girl that she has not priced herself out of the marriage market by choosing a career. This last point is especially effective if the mother is also pursuing a career (or at least working) and the marital relationship is good, providing a model for a successful career-marriage combination. Other men with favorable attitudes, such as teachers, can also play this role, but the importance of exposure to men married to career women cannot be minimized. Thus Alice Rossi maintains, "Young women need exposure to women scientists and doctors as models of what they might themselves aspire toward, but they also need exposure to men married to such women, as models of the kind of husbands they might have the courage to seek for themselves."[50]

Evidence from several studies supports the role of paternal attitude. In Lopate's study of women medical students, fathers were seen as especially instrumental in supporting the decision to study medicine (also, a higher proportion of women medical students had parents who were physicians than did their male colleages). Rossi reports that women in a national survey viewed fathers as only slightly more disapproving of careers for women than mothers, but much *more* approving than husbands (or close male friends). Rossi notes that it is much easier for a father to be proud of his daughter's achievements than for a husband to live with the consequences of his wife's (and this paradox is apparently perceived as well by women).

Thus, encouraging career choice is not just a matter of changing women's attitudes but those of men as well. Because of the generation gap, girls are probably first exposed to somewhat out-of-date male attitudes through their fathers (although basic male—or female – attitudes have not changed as fast as one might think). Thus Rossi points out that "Some women are no doubt hiding behind the rationalization that they approve of certain career goals for women but do not pursue such goals themselves because men disapprove of women with higher career aspirations and men are more important in their lives than careers are. Yet it is important to point out that such rationalizations are at work among women, and to indicate that men's attitudes must be changed as well as women's, if more women are to take professional careers seriously in American society."[51]

3. *Social class.* Parental education and occupational level are extremely important as general determinants of career choice for both men and women. For women, this factor may be especailly critical. The idea of higher education for women is predominantly a middle- and upper-class value, and the ideal of the traditional status of women is more attenuated in these classes. The first pioneers in careers for women were exclusively from the upper and (predominantly) upper-middle classes. Social class is itself a determinant of whether mothers will have followed a career as well as of socialization practices, so that social class is really a global factor affecting many others mentioned here. It is important to note, however, that Friedan's "feminine mystique" is described as a peculiarly middle-class phenomenon, especially of the suburbs, where women of some education (usually college) are discouraged from entering careers. Thus, although social class is itself a factor, differences *within* social classes are also great, so that many of the other factors mentioned here are only very imperfectly linked to social class.

All major studies of women pursuing professional careers show that social class is important, although some do not have a control group for comparison. One that does is Lopate's study of medical students, in which parental financial, educational, and occupational status was higher for women than for men. Williams reports that Radcliffe graduates who went on to medical school were likely to have fathers in the professions or high business and management posts (a total of 161 had such positions in the sample of 203). No control group of Radcliffe students is reported, but the high proportion who had fathers who were physicians (45 out of 203) or professionals (74 out of 203) as compared with businessmen (42 out of 203) is likely to be higher than in the college as a whole. One might predict that within the upper-middle class, businessmen would have more conservative ideas about women than would professionals. Ginzberg's study of graduate women found that about two-thirds of the fathers had held professional, academic, or executive positions, and that at least one parent (usually the father) of 60 percent of the women was a college graduate. In addition, Ginzberg found that about three-fifths of his sample were only children or had only one sibling, so that family size (a social-class-linked characteristic) might also be an important variable.

4. *Peer group.* Aside from early family influences, peer group is likely to make an independent contribution. This effect is especially strong as it affects the level of aspiration of women in school. The inhibiting effect of the presence of men on women's scholastic striving is one common reason advanced for separate women's colleges, but this sex-segregation may have the opposite effect and plant the mystique even more firmly. Intelligent girls (and boys) are likely to be put into honors classes or tracks (or even special schools, e.g., Bronx High School of Science in New York) so that the level of peer group aspiration and pedagogical encouragement is much higher than that of the general school population. Such settings may be particularly important in encouraging women's career choices, especially if family influences have not been very great, since the societal pressure for career entry is much less (or even negative) for women as opposed to men. The resulting social network from early peer group contacts is likely to have reinforcing effects on later career decisions.

5. *Region of country and rural-urban dimension.* One global factor of importance may be region of origin. For example, the cult of Southern womanhood is likely to be particularly antagonistic to career choice. No data on regions as a variable were available, but after suitable controls for average regional educational level are applied, one might expect, for example, a lower proportion of women medical students from the South than from the Northeast or Far West. The rural-urban dimension has been found to be important in two studies. Lopate reports that 60 percent of women medical students come from large cities or suburban areas whereas only 50 percent of the men did so. In the Ginzberg study, 75 percent grew up in urban areas (no comparable male or female-noncareer group is reported), and he notes "Most of these girls had spent their formative years in an environment where new trends affecting women, including the new pattern of women working out of the home were accepted."[52]

6. *Demographic pressures.* This factor has not been as important in the United States as in other countries, notably the Soviet Union. What "demographic pressures" means is "shortage of men." In this country this has occurred during wartime, especially

during World War II, when women were selectively encouraged to attend medical and professional schools and to work outside the home in war-related industries (which sometimes led to more enduring career interests). Thus Rossi notes:

> Changes in the status of women and of the relations between the sexes tend to occur when they suit the needs of society, which often means when they suit the needs of men. The traditional view of woman's role is always shelved for the duration of a war, during which women are praised for leaving their homes and holding down a "man's job" while men are in the military services. At the end of both world wars, fewer women have returned to traditional lives than had left their homes at the beginning of hostilities.[53]

Nowhere has the effect of demographic variables had more dramatic effect on women's career choices than in the Soviet Union. Dodge[54] notes that the sex ratio (number of males per one hundred females) dropped from ninety-nine in 1897 to ninety-two in 1939 to *seventy-four* in 1946 and has risen slowly to eighty-three in 1959 and to a projected ninety-two in 1980. The devasting effect of World War II, as well as political purges, resulted in a tremendous influx of women into occupations at all levels, encouraged by explicit government policy. Also contributing to this influx was the high number of unmarried women as a natural result of the shortage of men. Thus, approximately 75 percent of all physicians are wommen, as are 53 percent of all professionals, especially scientists and engineers. Most (80 percent) of the employed women, however, are still engaged in heavy, unskilled work. Despite the high proportion of women professionals, their rate of advancement and level of achievement is still behind that of men. Dodge attributes this to the marriage-career conflict with its related problems of childbearing, child rearing (child care facilities are much more extensive than in the U.S., but are not universal), and lack of geographic mobility attendant upon remaining with the husband. Recently, with the rising sex ratio, some relative discouragement of female entry into careers has been noted. We will specifically address the Russian case in the final chapter.

Although the above examples involve demographic pressures resulting from war, clearly demographic pressures exist right now in this country that could propel women into careers. There is the perennial shortage of physicians and scientists, and women are invariably lauded as a great "untapped resource" for such positions. Yet the campaigns to encourage women to enter these professions have been less than spectacular, and Rossi maintains they are doomed to failure:

> Campaigns to increase the support and encouragement given to the *college-age* woman to enter the sciences, engineering and medicine can only effectively reach and help the young woman who is already interested and prepared by a background in science and mathematics to take advantage of opportunities offered her in college. Such women are a tiny minority of their sex, whose experiences at much earlier ages have set in motion an abiding interest in things generally disapproved of for girls in our society. College freshmen do not shift from fine arts to chemistry, or from journalism to engineering, except in rare instances. Hence efforts to be really effective must concentrate on much earlier stages of life and *must involve fundamental changes in the rearing of girls and boys.*[55] (Italics added.)

Thus the problem of women's career choice cuts much deeper than could be remedied by having a few television commercials featuring women scientists (though that would help, too). It involves the key variable of *socialization,* which bridges the gap between the social structural status and image of women and their modal personality characteristics. The life plan of women is the variable of concern and it is the domestic value complex that has for a long time determined this life plan.

7. *"Real" problems.* If the complex of domestic values were to vanish tomorrow, there would still be substantial social factors impeding women's career choice. In dual career marriages, two positions must be found in the same geographic area, not always an easy task. There are very real problems in combining motherhood and a career, such as the high cost and low availability of qualified and reliable child care and household help. We have already discussed this issue in an earlier section. For those who wish to spend the first years at home with their children, there should be an "institutionalization of flexibility" built into all careers so that women are not penalized for wanting to combine marriage and career. Thus, Lopate points out that although male resident physicians are readily granted two years' leave for military service, maternity leave for women residents is severely frowned upon.

Discrimination against women in professional fields may be so deeply a part of the professions as to be almost functionally autonomous of the continuation of the values of the mystique. The increased influx of women into careers should not blind us to the fact that many of these choices are still "compromise careers," based on the perception (sometimes, but not always, exaggerated) of real discrimination in the field. Thus, as careers open up to women, certain sub-areas (usually less desirable and remunerative) are likely to be marked off as "women's work," while others remain relatively closed to women. For example, a woman lawyer may find few barriers to entering estate planning or tax law and may of necessity settle for these fields even if her strongest interests lie in more closed areas, such as criminal or corporation law. In my career study, one degree candidate in the education of disturbed children describes her compromise as follows:

> My long-standing interest in newspaper work was put aside partly because of the heavy competition with males that such work involves. Also women journalists tend to get stereotyped work assignments—heavy on society and cooking, light on politics and such. So I took a second choice where I could be sure of immediate personal rewards and continuing intellectual growth without beating my head against the wall of competition.

Thus the continued predominance of women in "women's fields" should not be taken as proof of a peculiar personality congruence to such roles, as some contend, but may represent women's second choices in response to very real discrimination.

Concerted action against discrimination will be necessary (as it is in racial discrimination) to bring down the barriers against women. The laws do exist (all major civil rights legislation specifies sex as well as race), but may be difficult to enforce. Already, subtle progress has been made. For example, the "Help Wanted" sections of newspapers are now consolidated so that there are no longer separate listings for "Help Wanted—Male" and "Help Wanted—Female." Although this step will probably have a negligible effect on hiring practices, it may have other, more lasting effects. For example, who can tell what it will mean for aspiring girls to leaf through the Sunday papers and see the columns of ads for biologists, chemists, physicists, financial analysts, lawyers, engineers, etc., listed under "Help Wanted—Male/Female" rather than hopelessly poring through the "Help Wanted—Female" columns for a pitiful few two-sentence ads for biologists (usually in food industries and sounding suspiciously like lab technician jobs) buried among the avalanche of enticing ads for secretaries, nurses, clericals, typists, etc. (and little else), while the inviting pages of ads for scientists, economists, etc., lie in solitary splendor under the forbidding heading of "Help Wanted—Male." It is of such small stuff that social change is made.

THE MALE CAREER AND FAMILY MODEL

While a potential choice exists for many women as to their career goals, this choice is not open to men. Even when women must work because of economic necessity, they do have a choice as to whether they will be defined by their work or family roles. Even if they wish the former, they often find themselves squeezed into the latter by society's view of their role. In any case, the common societal pattern is for work to confer identity on the male, but not on the female. This pattern doubtless stems from the values of domesticity being primary for the female, while the function of economic support has been culturally assigned to the male, even if in reality the female is contributing a large part of the financial sustenance to the family group. An extension of the primacy of this economic support function has been the cultural emphasis on male achievement and typing of males by occupation.

The life plans of men, therefore, often are shaped by their vocational goals, especially in the case of professional careers. An early emphasis on achievement and competition and "getting ahead" becomes meaningful when considered in the context of this life plan, where work plays the central role. Recent critiques of the male sex role, such as Marc Feigen Fasteau's *The Male Machine,*[56] emphasize the negative aspects of these constraints, in which the male is expected to be task-oriented and competitive at the expense of his emotions. The rising incidence of stress-related diseases among men, such as heart attacks and ulcers, has been cited as a cost of the push to achievement.

Society provides a male subculture of values and interests to match the female model of domesticity. In our society, this value system is somewhat less obvious than the female, perhaps because the latter has been so often presented in advertising and the media. Nonetheless, some images emerge. Most emphasize the competitive, achievement-oriented stance, sometimes with a negative implication for familial involvement. The images of the athlete and (to a lesser degree nowadays) the military hero come to mind. Interest in sexuality per se outside of the constraints of the family system, as epitomized by *Playboy,* for example, is also part of the cultural male image. Historically, the figures of the cowboy and the pioneer have had a strong hold on male imagination, perhaps because of the combination of aggressive, achievement-oriented values with an implied denigration of the (weaker) female sex and the constraints of domesticity. We shall dwell more on these images in the next chapter on symbolic views of sex roles.

To what degree is the greater male propensity to aggressiveness at the root of the male achievement system? We have seen the considerable weight of biological evidence presented on the higher incidence of aggression in men; of what relevance is this factor for male life plans? There is probably an underlying component of aggression in the male value system, which serves to energize the achievement motive. Is this tantamount to saying that the life plans of men (and women) are determined by biological factors? No, because the achievement goal is to be sought after within the constraints of a life plan largely determined by the nature of the family, and especially the (relatively) new ideals of domesticity. This is not to say that the aggressive component is not there, but rather that it only achieves meaning within a social system that prescribes occupational achievement and self-definition for men but not for women. The origins of this prescription are considerably more complex than an extra amount of androgen would explain. While the nature of the family (insofar as it concerns women's bearing children) is biologically determined, the value system of domesticity and definition of female identity totally in terms of that system are not. It is the latter two factors that determine the female life plan rather than the former, on which it is loosely structured.

What of the familial roles of men, that is, those of husband and father? They do, of course, play a role in the life plans of men, but not a defining one, as in the case of women's family roles. Symbolically, we see the difference in marriage when the husband's name remains unchanged but the wife (usually) changes her name to that of her husband. In English common law, Blackstone declared in 1760: "By marriage, the husband and wife are one person in law; that is, the very being or legal existence of the woman is suspended during the marriage, or at least is incorporated and consolidated."[57] Until recently, the legal structure upheld this convention by (among other things) requiring the husband's permission for the wife's business transactions and other matters. Since economic rights have primacy before the law, the male carrier of those rights was assigned the superior role in public matters.

For men, the role of husband is to a large degree an auxiliary one to that of breadwinner. Even when men are found in low-status occupations that confer little positive identity on their occupants, the link to the family system remains secondary. Taking the place of occupational identity is often involvement with the peer group of male buddies and the symbolic world of sports. Although many men may be very involved with their families, their primary social identity and self-identity does not come from this realm as it does for women. Whatever women may achieve in the public occupational world, their primary social and (often) self-identity come from the private, domestic one. Even single women are more often defined by their marital status than single men. Men, regardless of their involvement with the family, are often seen by others (and by themselves) as only temporary visitors to domesticity, while their prime involvement is in the external world of male culture.

Recently more attention has been given to the importance of domestic involvement for men, both as an escape from the pressures of the business world and as a necessity for the smooth functioning of marriage and the family. Participation of males in housework and child care is no longer rare and is not commonly taken to be a sign of "softness" and undesirable feminine tendencies as it once was. Several popular and professional books have appeared on fatherhood and its importance in the life of the child.[58] However, both housework and child care are commonly viewed as secondary activities for men and as aids to the primary housekeeper and child-rearer, the wife and mother. A man "helps" his wife with "her" housework and child care, even if they both work. The alternate phrasing, of a woman helping her husband with his household duties, does not sound right—it does not accord with our traditions of sex role division of labor and responsibilities. In the final chapter we shall see how critical this mental set is to difficulties in sex role change.

The tendency of a man not to identify with his home and the coupling of his identity with that of his job has led to many difficulties. Wives and children often lead secluded lives, given the long absences of men on the job. Now that few men work in the home, the separation of home and workplace imposes a greater separation between women and men. Such separation (and non-involvement when the man is home) imposes considerable stress on marriage and family ties. The stress of the job often overtakes men, resulting in disease and early death. When men do survive until retirement, they often find that they are ill-equipped to carve a life outside of work and fall prey to boredom and a sense of meaninglessness. In any case, the division of female domesticity and male occupational involvement presents any number of psychological and practical difficulties, although the system does allow for a meshing of roles when it is working optimally. Unfortunately, both conventional roles take their toll on their occupants. The female role in particular is getting stretched beyond recognition with the increasing number of married women working outside the home who are yet still largely defined by their role within it.

FINAL THOUGHTS ON THE FAMILY

The psychological and social concept of separate sexual spheres within the family is primary to our understanding of the sex role system. From early childhood, the life plans of men and women are shaped by their expectations of adult roles to be enacted within the constraints of the nuclear family system. Increasing trends toward female employment and lower birth rates have changed the supporting structure of the system, in that often women are no longer literally in the home. But they are psychologically in the home just as surely as if they were physically present. They see themselves and are seen by others as wives and mothers first, workers second. This secondary definition of work in the female life plan helps determine the choice of work in lower-level occupations that do not require much training or commitment. Women go into such jobs because they are not pushed to do otherwise as men are. At an early formative age, they perceive work as a secondary option, less important than family goals in their life plans, and therefore do not prepare themselves for serious careers. Of course, real-world discrimination and difficulties also stand in the way of such achievement.

Large-scale female participation in the upper levels of the work force is critical to change in the sex role balance of power system. On an individual level many women may choose not to work or to work only at secondary levels, and they may have made the right choice for themselves. For social change, however, something more radical is required. And, of course, the real problems of child care and housekeeping will not be solved until they become men's problems as well, for otherwise women assume a dual burden. Women must be relieved of exclusive responsibility for the domestic sphere if they are to participate fully in the public one. As we shall see in the final chapter, this problem has been a major stumbling block to sex role change.

We turn now to a consideration of the symbolic system of sex roles, that shadowy world of taboo, witchcraft, myth and ritual, language, literature, and the arts, which so closely follows the patterning of the societal roles of women and men.

SUGGESTED READINGS

Myron Brenton. *The American Male*. Greenwich, Conn.: Fawcett, 1966 (p). Written before the present concern with male sex roles; some perceptive insights on the "masculinity trap."

W. Elliot Brownlee and Mary M. Brownlee, eds. *Women in the American Economy: A Documentary History, 1675 to 1929*. New Haven: Yale University Press, 1976 (p). Documents, public and private, on working women.

Susan Brownmiller. *Against Our Will: Men, Women and Rape*. New York: Simon & Schuster, 1975 (p). Rape as integral to male-female relationships and the balance of power between the sexes: a feminist view.

Ann Cornelisen. *Women of the Shadows*. Boston: Little, Brown and Co., 1976. An excellent and poetic account of female peasant life in southern Italy.

Rose Laub Coser, ed. *The Family: Its Structures and Functions*. New York: St. Martin's Press, 1974 (p). Many classic articles with special attention paid to the role of women in the family.

Cynthia Fuchs Epstein. *Woman's Place: Options and Limits in Professional Careers*. Berkeley: University of California Press, 1970 (p). Problems of women in careers.

Marc Feigen Fasteau. *The Male Machine*. New York: McGraw-Hill, 1974 (p). A critical look at what "masculinity" entails.

Betty Friedan. *The Feminine Mystique*. New York: Dell Books, 1963 (p). Still the best discussion of the ideology that is at the basis of woman's place in America.

Ernestine Friedl. *Women and Men: An Anthropologist's View*. New York: Holt, Rinehart & Winston, 1975 (p). Concise treatment of sex roles in hunting and gathering and horticultural societies.

Lois Wladis Hoffman and F. Ivan Nye. *Working Mothers*. San Francisco: Jossey-Bass, 1974. The consequences for wife, husband, and children.

Louise Kapp Howe, ed. *The Future of the Family*. New York: Simon & Schuster, 1972 (p). Readable articles on the American family highlighting sex role changes.

Leo Kanowitz. *Women and the Law: The Unfinished Revolution*. Albuquerque: University of New Mexico Press, 1969 (p). How the law codifies sex roles and how new legislation could affect the system.

Mirra Komarovsky. *Blue-Collar Marriage*. New York: Random House, Vintage Books, 1967 (p). Classic study of the lives of workers emphasizing the separate spheres of women and men.

Mirra Komarovsky. *Dilemmas of Masculinity: A Study of College Youth*. New York: Norton, 1976. A study with many findings relevant to sex roles.

Juanita M. Kreps, ed. *Women and the American Economy: A Look to the 1980s*. Englewood Cliffs, N.J.: Prentice-Hall, 1976 (p). Excellent essays integrating economic and sociological views.

Ruth B. Kundsin. *Women and Success: The Anatomy of Achievement*. New York: William Morrow, 1974 (p). Profiles of successful professional women.

Joyce A. Ladner. *Tomorrow's Tomorrow: The Black Woman*. New York: Doubleday, Anchor Press, 1972 (p). Study of the black woman's role in family and in society.

Gerda Lerner. *Black Women in White America: A Documentary History*. New York: Pantheon Books, 1972 (p). Well-done anthology on the situation of black women in the United States.

Cynthia B. Lloyd, ed. *Sex, Discrimination, and the Division of Labor*. New York: Columbia University Press, 1975 (p). Sophisticated economic thinking brought to bear on the issue of sex roles.

David B. Lynn. *The Father: His Role in Child Development*. Belmont, Calif.: Wadsworth Publishing Co., 1974 (p). The forgotten parent and his effect on his children.

Lucy Mair. *Marriage*. Baltimore: Penguin Books, 1971 (p). Anthropological perspective on the role of marriage in the social system. Clear presentation of Mair's own views and those of others.

M. Kay Martin and Barbara Voorhies. *Female of the Species*. New York: Columbia University Press, 1975 (p). Good treatment of "woman's place" in various societies, from the hunting and gathering types to the industrial.

Bernard I. Murstein. *Love, Sex and Marriage Through the Ages*. New York: Springer, 1974. Interesting historical approach.

Ann Oakley. *Woman's Work: The Housewife Past and Present*. New York: Pantheon, 1974 (p). The history of woman's primary occupation.

Lee Rainwater, Richard P. Coleman, and Gerald Handel. *Workingman's Wife: Her Personality, World and Life Style*. New York: MacFadden-Bartell, 1968 (p). The feminine mystique among the working class.

Rhona and Robert Rapoport. *Dual-Career Families*. Baltimore: Penguin Books, 1971 (p). Case studies from Britain; good insights into the subtle workings of sex roles.

Rayna R. Reiter, ed. *Toward an Anthropology of Women*. New York: Monthly Review Press, 1975. Well-done compilation of theoretical and empirical work on the nature of the family and women's roles in widely disparate societies.

Michelle Z. Rosaldo and Louise Lamphere, eds. *Woman, Culture and Society*. Stanford: Stanford University Press, 1974 (p). Provocative theoretical pieces integrating much current anthropological thinking on the subject.

Ronald V. Sampson. *The Psychology of Power*. New York: Random House, Vintage Books, 1968 (p). Women and power in the family.

Edward Shorter. *The Making of the Modern Family*. New York: Basic Books, 1975. Psychohistorical theory of change in feelings as the basis for modern family life and domestic values.

Robert W. Smuts. *Women and Work in America*. New York: Schocken Books, 1971 (p). The classic work on this subject.

Joyce Teitz. *What's a Nice Girl Like You Doing in a Place Like This?* New York: Coward, McCann & Geoghegan, 1972. Eleven professional women tell their stories. Some good insights into how the sex role system operates.

Athena Theodore, ed. *The Professional Woman*. Cambridge, Mass.: Schenkman Publishing Co., 1971 (p). Valuable collection of articles about women in a variety of professions.

4

The Social Maintenance System: Symbolism

In addition to more or less direct inputs from the biological, psychological, and social systems, the contribution of the symbolic realm to the maintenance of sex roles cannot be underestimated. When we use the term *symbolic,* we are referring to those social representations of the relationship between the sexes that tend to reflect existing social conditions as well as to aid in affirming and maintaining them. We will specifically deal with several general symbolic images: the concept of sexual pollution (especially in reference to the menstrual taboo), witchcraft and the general problem of evil, mythic themes in sexual representation, and initiation rites and other rituals. We will pay special attention to particularly articulated systems of representing sexuality and sex roles, such as the Chinese Tao and the Indian Tantra. Wherever possible we shall emphasize the impact of Western religious traditions, Judaic and Christian, in shaping our view of the sexes and our sex role system. Finally, we shall look at specific examples of symbolic usage in language, literature, and the arts.

Throughout this chapter we shall be looking at symbolism as it reflects sex roles and as it helps maintain them. In the first case, we shall see that symbolic systems reflect the social divisions in the societal structure. For example, the belief in witchcraft tends to flourish in societies with deep structural divisions between the sexes that emphasize the subordinate position of women (who are the ones usually accused of witchcraft). In that sense, we can think of the belief system as merely a reaffirmation of existing social structures. However, at the same time, these symbolic systems serve a socializing function in that they inculcate beliefs that help keep the existing sex role system in smooth operation. They also act to reward and sanction behavior in accordance with the needs of the system. For example, in the case of witchcraft, we can see that female fear of incurring accusations of witchcraft would serve to discourage deviant behaviors that might serve as the occasion for such accusations. A more contemporary example would be the cinematic image of women, which tends to reaffirm the values of beauty and romance in female life plans.

Along with the twin reflective-maintaining character of symbolic representations comes the need to consider simultaneously both psychological and sociological factors in the understanding of such representations. Although we are classifying symbolism under the sociological rubric because it is a collective representation, it obviously exerts its power through the psychological system. Thus to understand the menstrual taboo, we must consider both its impact on individual attitudes and feelings about women and sexuality and its articulation with the more general social system of sex roles. Franz Steiner, in his classic study on taboo, isolates these two approaches and emphasizes their complementarity:

> The narrowing down and localization of danger is the function of taboo of which we are now speaking. The dangerous situation is then defined in terms of such localization, which in turn is meaningless without abstentive behaviour. *It is the job of the psychologist to study the emotions of fear in terms of the human mind,* to conjecture the situations in which these fears are allayed, and to relate these situations to the conditioning of the individuals concerned. *But to study how danger is localized in social institutions,* and what social pressure is needed in order to regulate abstentions so that the danger can remain localized, *is to approach the problem sociologically.*[1] (Italics added.)

Throughout this book, we have stressed the need to take a truly social psychological perspective—in which both approaches are considered at once—to clarify the nature of sex role maintenance. In the area of symbolic representations, we have an ideal case for the application of this method. Historically, analyses of symbolism have tended to emphasize one or the other approach. Raymond Firth,[2] the anthropologist, has singled out Sigmund Freud and Emile Durkheim as the prime proponents of the psychological and sociological approach, respectively. Freud's position is, of course, well known. As many have put it, society is the individual writ large. Social structures are merely projections of individual needs and conflicts and can be understood as such. We shall see the prime example of this sort of thinking in the Freudian analysis of the menstrual taboo in terms of castration anxiety and the Oedipal complex. Durkheim, the great sociologist, on the other hand, sees symbols as the affirmation of the individual's tie to society and as tools for maintaining that social feeling of solidarity and common purpose that in the end defines social structure. Religious ritual, for example, is seen as an embodiment of social sentiment where the ultimate object of worship is the society itself. We shall see that both approaches will be useful in understanding the role of symbolism in the sex role system.

SEXUAL POLLUTION AND THE MENSTRUAL TABOO

One of the most widespread taboos is that which prohibits contact with a menstruating woman.[3] Cultures vary in the severity of the taboo. For some it concerns mainly restrictions against sexual intercourse. For example, orthodox Jews practice *niddah* or separation during the menstural period and for seven days after it. Immersion in a ritual bath (*mikvah*) is required before intercourse can be resumed. The woman is referred to as "ritually impure" during this time, an appellation also applied to the woman after childbirth and to various prohibited foods and practices. (Interestingly enough, in Judaic law, a woman remains unclean for a longer time after the birth of a girl than of a boy.) The practical outcome for cultures that practice such menstrual sexual restrictions is to encourage intercourse during the woman's most fertile period, thereby building up sexual desire and maximizing the chance for successful conception. But recourse to a population-building explanation is too rationalistic a solution for such a subtle problem. Many cultures that prohibit intercourse during menstruation do not stop at this, but generalize the attitude of fear of contamination to other areas. Sometimes menstruating women are prohibited from contact with food (especially food eaten by men) or with cultivated fields, in the fear that some harm will come to the living substances touched by the menstruating woman.

Why is the menstruating woman such a danger, especially to men? Most societies' attitudes range between images of impurity or uncleanness and outright danger. The justification behind the fear of the unclean state is usually the expectation that some evil (sometimes specified, sometimes not) will befall the breaker of the taboo, and most of all, the person or object touched by the forbidden woman. Lurking behind this image is a definite sense of power in menstruation itself. Margaret Mead and some other anthropologists have interpreted the taboo as bespeaking a nearly universal fear of blood, especially the "magical" menstrual blood, which seems to appear and disappear at regular intervals without harming the woman. This blood, then, is special, perhaps magical, since when men bleed they are hurt and often die. In the absence of any scientific explanation for menstruation, the magical one would gain credence. Yet why should this "magic" evoke fear and specifically fears of danger to men? Why should so many cultures come to the same conclusion about this phenomenon?

The explanation of fear of blood is not very persuasive. In a sense, it just restates the problem. Why should this blood evoke fear? After all, the menstrual taboo is really a blood taboo, so we have just renamed the phenomenon. Also troubling is the fact that many cultures that have rigid menstrual taboos, such as the Mae Enga of New Guinea (referred to earlier, see p. 30), also have equally strong fears of sexual pollution from intercourse. In almost all cases this fear is stronger on the part of men than of women, though, presumably, the act of intercourse is an "equivalent" one on both sides. Certain tribes have very specific prohibitions against intercourse before warfare, for example, reflecting the fear that the men will be weakened and jinxed in battle. Needless to say, contact with a menstruating woman is considered to be even more dangerous.

The menstrual taboo, then, seems to be generalizable as a more widespread fear of the power of women. When we turn to witchcraft we shall see this familiar theme reemerging and shall treat it at greater length. For the time being, let us turn to more specific psychological and social explanations for the menstrual taboo. The Freudian view of the menstrual taboo concerns itself with castration anxiety. If we recall our earlier discussion of the Oedipal period and sex role development, we see that the male's view of the female genitalia is that of a wound, the aftermath of castration. According to Freud, women also have this view and thereafter mourn the loss of their penis through a lifelong devotion to penis envy. In any case, menstruation is seen as an experience that emphasizes the idea of the female genitalia as a wound (it is bleeding, after all) thereby reevoking fears of castration among men and rage against their anatomical fate among women. Women are angry during this time because of their reawakened trauma and men fear the Oedipal punishment once again: the result is the menstrual taboo. Not a bad try at an explanation, certainly a bit deeper and more convincing than the fear-of-blood argument. Yet belief in it requires prior belief in Freud's psychosexual conflicts, as well as belief in their lifelong effects and power to affect the social structure. For a firm believer in the Freudian system, such steps are easy and natural ones. For the more skeptical observer, another perspective may be sought. This perspective is the sociological.

Recall Steiner's comment (p. 159) on the need for a dual approach to taboo. One can understand fear in terms of the human mind (as Freud has attempted to do), but this does not preclude the need to seek the complementary sociological perspective. Why is danger seen as emanating from a specific sector of the population, women, at a specific time, menstruation? For an answer, we must look at the structures of societies that hold such beliefs. The British anthropologist Mary Douglas[4] has taken such an approach. In her view, the source of tabooed danger in a society is symptomatic of a source of unarticulated power. The society itself is often structured in an ambiguous or contradictory way, so that there is reason to fear unknown dangers, especially from women, who lack any source of legitimate power and hence must resort to magical means. In Douglas's view, then, fear of the menstruating woman and of witchcraft come from the same source. In fact, witches are sometimes associated with menstruation in legend. Douglas cites the case of the Mae Enga of New Guinea, where notions of menstrual taboo and sexual pollution run rife. She notes that in this tribe the men marry the daughters of their traditional enemies, and hence "the marital relation has to bear the tensions of the strong competitive belief system. The Enga belief about sex pollution suggests that sexual relations take on the character of a conflict between enemies in which the man sees himself as endangered by his sexual partner, the intrusive member of the enemy clan."

WITCHCRAFT

Few topics in anthropology have aroused as much interest as witchcraft, and theories abound as to its origins and functions. Students of European and American history have also speculated widely about this topic, discussing the outbreak of witchcraft trials in Europe in the sixteenth and seventeenth centuries and in Salem during colonial times. Some very broad parallels can be drawn between African and European witchcraft (Africa being the area of the world where witchcraft ideas are most prominent). The accused are almost always women; there is usually an idea of flight to a witches' assembly; and the victims are often children or persons afflicted with wasting diseases thought to be caused by witches' blood-sucking. The European variety of witchcraft was complete with an elaborate demonology in which the witches' sexual intercourse with the devil or with other demons often played a prominent role.

Why are witches usually women? Remy in a 1595 treatise commented:

> Certainly I remember to have heard of far more cases of women than men: and it is not unreasonable that this scum of humanity should be chiefly drawn from the feminine sex . . . since that sex is the more susceptible to evil counsels.[5]

Here we again meet the notion of woman as purveyor of evil, a theme suggested by our discussion of the menstrual taboo and ideas of sexual pollution from women. The intersection of cultural views of women and other symbolic ideas with the need for an explanation for seemingly inexplicable phenomena seems to account for most instances of "witchcraft." In Africa and medieval Europe, cases of infant death and adult sickness cried out for an explanation in human terms. In Europe the additional input from Christian ideas of demonology gave a particular flavor to witchcraft explanations. But, to return to our original question, why was the answer phrased in terms of the evil *woman?*

Anthropological thought on the question has tended to dwell on the subordinate position of women as the source of the accusations. On one hand, it is safe to accuse the powerless; they become the scapegoats of the society. On the other hand, the very lack of power may in itself arouse suspicion. Goody writes:

> It seems likely that because of their basic identification with domestic and kinship roles, women will usually be denied the legitimate expression of aggressive impulses. But in whichever roles aggression is not legitimate, it is in these that we can expect to find that imputations of covert mystical aggression are made, and, further, that they evoke publicly sanctioned counteraction; that they are considered evil.[6]

The male authority system of legitimate power evokes one kind of response of fear, the female "non-system" evokes another, but one that is less defined. Mary Douglas speculates:

> . . . where the social system is well-articulated, I look for articulate powers vested in the points of authority; where the social system is ill-articulated, I look for inarticulate powers vested in those who are a source of disorder. I am suggesting that the contrast between form and surrounding non-form accounts for the distribution of symbolic and psychic powers: external symbolism upholds the explicit social structure and internal, unformed psychic powers threaten it from the non-structure[7].

For Douglas, the seemingly "powerless" segment of society will be seen as the source of invisible "powers." When women hold culturally ambiguous roles, they are more likely to be seen as the source of evil.

Witchcraft accusations against women can be seen as part of a wider phenomenon of ascribing the origin of evil to women. The Pandora's box myth comes to mind, as does the idea of original sin in the Garden of Eden. In the latter case, Adam's fall is ascribed to his yielding to the temptation to eat the apple offered by Eve. Many other stories tell of temptation emanating from woman and leading to the downfall of man. The legend of the Sirens luring sailors to their death is such an example. Often, especially in a Christian context, the idea of sexual temptation is an integral part of the negative image of woman. The prominent place given to pornographic portrayals of witches' intercourse with devils in Inquisition trials of witches testifies to the existence of heavy sexual repression and the consequent denunciation of those seen as the origin of the forbidden impulse: women. Often there seems to be a one-dimensional view of woman: she is her sexuality, no more, no less. When that sexuality is seen as threatening or evil, so is she. When a more positive view of her sex is expounded, she is elevated to a more favorable position, but still a one-dimensional one. In our own society, to a limited extent, the latter instance is the case, with female sexuality being a major ingredient in advertising and other media images of woman.

Persecution of witches (and the more general idea of female evil as well) can be seen as a mechanism of social control, as an integral part of the system of male power over women. Spooner suggests this in his discussion of the "evil eye" (usually possessed, of course, by women) when he notes the tendency for women to be accused because of their more tightly defined role:

> Women are particularly liable to suspicion since their social role is more strictly defined than men's, and they are at a physical and social disadvantage to start with, any unusual behaviour, or any trait that prevents them from fulfilling their women's function, may make them suspect: e.g., barrenness, brashness, unexplained visits, etc.[8]

Presumably, women would be discouraged from such out-of-role behaviors by their fear of being accused of witchcraft as well as by more immediate sanctions. Parrinder concludes: "The subjection of women was one aim of the witch-hunting,"[9] and notes that many male secret societies in Africa have witch-hunting as one of their primary reasons for being. The woman who had no particular social niche, especially none in relation to a male (which is the usual way of defining a woman's status) was particularly likely to be accused. Parrinder notes:

> Many of the accused were women whose position in society was uncertain, who had outlived their husbands, and lived themselves in poverty and were suspect by their very isolation. Their minds would naturally harbour (or be suspected of harbouring) jealous thoughts against those more fortunate than themselves, who had children and security. The very suggestion of an old woman approaching a child, in affection or for sympathy, would send a shudder through the breasts of the well-off.[10]

Also, after the child-rearing years, women often lacked a well-defined social function, so their position was quite ambiguous. Recall Douglas's analysis of the powers imputed to socially ill-defined persons:

> Where the social system requires people to hold dangerously am-

biguous roles, these persons are credited with uncontrolled, unconscious, dangerous, disapproved powers—such as witchcraft and the evil eye.[11]

It is clear that a complex phenomenon such as belief in witchcraft has many causes. It grows out of the complexities of social structure and the human mind's search for a comprehensible explanation for incomprehensible events, such as death and illness. In the European case, pagan superstitions from ancient Europe provided a framework into which Christian demonology and notions of sexual temptation and evils could fit. That this framework seems to emphasize the role of women as witches almost universally can perhaps be traced to their subordinate position, which makes such accusations "safe"—and likely, given the ill-defined place of women in the formal social structure. Women (and sometimes men) who experienced delusions and strange dreams were identified as witches rather than as the mentally ill because of the preexisting societal belief in witchcraft. One recent analysis of the Salem witch hunt suggests that ergot poisoning (from rye flour poorly stored) may have caused at least some of the delusional behavior reported during that siege. But ergot poisoning alone could not have created Salem; the events had to be fitted into an already existing cultural system of belief in female witchcraft, a system that had a dual existence in the social structure and in the minds of its citizens.

We see, then, a continuity of thought between ideas of menstrual and sexual pollution and of witchcraft. In both cases, the source of danger and evil is usually thought to be female. There are exceptions, of course. For example, Faithorn[12] has studied a tribe in New Guinea in which semen is considered to be a powerful substance, like menstrual blood, that must be kept under tight control. In such a society, women as well as men are affected by notions of sexual pollution. As we shall see in the following section, Eastern, particularly Indian, ideas of sexuality emphasize the weakening effects of ejaculation, but such images do not usually have the restrictive character of the menstrual taboo. The special quality of the menstrual taboo and notions of sexual pollution as they apply to women seems to be the tendency for the idea to generalize to a negative or dangerous image of women in general, rather than relating only to the sexual function. Such images, of course, jell with the prevailing cultural currents of thought about women as well as the actual social mechanisms that regulate women's lives. In the case of witchcraft, instances of male witches are not unknown, but they are rare. Accusations against males usually arise from specific kinship or economic rivalries and conflicts rather than out of their general position in society, as seems to be the case with women.

Menstrual taboos, ideas of sexual pollution, and witchcraft all reflect the social structure and the psychological image of woman arising from that structure. In addition, they help maintain that structure by inculcating a restricted and negative image of women—in the case of the menstrual taboo by defining them categorically and similarly on the basis of biological function. In the case of witchcraft, women's "proper" behaviors are reinforced by their fears of arousing witchcraft accusations. Such a system of beliefs can only serve on both the societal and individual levels to emphasize the structural power split between the sexes that led to the phenomena in the first place. By considering women categorically—as all alike—such beliefs further divide the society along sexual lines. In the individual psyche, the negative attitudes of men are further reinforced and developed by such beliefs. Young men growing up in such societies are given a distorted image of woman, which prevents its own disproof through the mechanism of the taboo and fears of evil. Women themselves are given a negative self-image that may serve to perpetuate their state, and in a sense, justify it. Mary Douglas's oddly poignant image of the witch comes to mind:

> Witchcraft, then, is found in the non-structure. Witches are social equivalents of beetles and spiders who live in the cracks of the walls and wainscoting. They attract the fears and dislikes which other ambiguities and contradictions attract in other thought structures, and the kind of powers attributed to them symbolize their ambiguous, inarticulate status.[13]

It is women who live in the "cracks" of the social structure and whose "inarticulate status" provides the occasion for the projection of all sorts of unspecified fears and terrors. Then this very symbolic system serves to maintain the social system that gave it birth and sustenance.

Our understanding of symbolic systems of sex roles will be incomplete, however, unless we consider some of the more general mythic and ritualistic themes of the sexes, to which we now turn.

SEXUAL THEMES IN MYTH AND RITUAL

We will consider myths and rituals together, although they in some sense constitute analytically separable categories. In the anthropological literature, a myth has been defined as a "sacred tale,"[14] a narrative representation of a metaphysical reality. For example, the numerous mythological accounts of creation are one manifestation of this genre. When we term a story a "myth," we do not pass judgment on its status as truth, that is, whether or not it "accurately" describes real events that happened in the past. For our purposes, questions of truth or falsehood are irrelevant. What does concern us is the social and psychological truth of these representations and what light they shed on the nature of the social system and psychological structures of which they are a part.

A ritual is a socially recognized behavioral enactment that is formalized in some way to mark a significant event in the life of the society and/or the individual. Rituals have a symbolic component that is very similar to myth, although they vary in their relationship to narrative accounts (the story-like feature of myths). Some theories of ritual maintain that all myths derived originally from rituals as narratives associated with rites.[15] In any case, the association is quite close in some cases and may be present in others where we lack historical material to flesh out our knowledge of lost rituals or myths. Mircea Eliade[16] gives an example of the mythical coupling of Heaven and Earth being reenacted as an agricultural fertility ritual by the ritual intercourse of the priest and his wife. The creation myth that lies behind the ritual is thought to confer creative fertility to the tilled earth.

Perhaps the most persistent symbolic sexual theme in myth and ritual is that of androgyny, of the idea of masculine and feminine principles or essences that together constitute the universe. In Western psychological thought, Carl Jung has adapted these ideas to the individual level. According to Jung, the male principle (animus) and the female principle (anima) are archetypes found in the collective unconscious of each individual. Jung developed the idea of the collective unconscious as an explanation for universally found themes in myths, rituals, and dreams. The collective unconscious has been called "the storehouse of latent memory traces inherited from man's ancestral past," and "a residue that accumulates as a consequence of repeated experiences over many generations."[17] The archetypes found in the collective unconscious represent predispositions that can emerge into consciousness and influence behavior. In the case of the animus and anima, each person has an internalized representation of the other sex. The anima has been described as "a personification of all feminine psychological tendencies in a man's psyche, such as vague feelings and moods, prophetic hunches, receptiveness to the irrational, capacity for personal love, feeling for nature, and—last but not least—his relation to the unconscious."[18] The animus, on the other hand, includes personifications of physical power, initiative and planned action, and meaning and spiritual profundity—presumably the more rational functions. Jungians believe that both principles must work harmoniously within each person to permit the greatest personal happiness and creativity.

Before considering the ramifications of these images of femininity and masculinity, let us consider several very similar religious and mystical systems that build on the common idea of contrasting and complementary feminine and masculine principles. Perhaps the best known to the Western reader is that of the Chinese Tao and the yin-yang principles. The common lay interpretation of this system is to suppose that the stereotyped traits of masculinity and femininity have been projected on the universe as a whole. When we look at the descriptions of yin and yang, we can see how this conception arose. In *I Ching,* the hexagram *k'un* (yin) for female is the passive principle. The commentary on it gives its essential flavor:

> Exalted indeed is the sublime Passive Principle! Gladly it receives the celestial force (of the Creative Principle) into itself, wherefrom all things receive their birth. . . . The mare symbolizes those (passive, female) creatures which wander throughout its confines. Gentle and accommodating, how auspicious is this omen!
>
> The Passive Principle, thanks to its exceeding softness, can act with tremendous power. Silent, tranquil, its virtue is amorphous until, receiving into itself the subjective force, it becomes clearly defined.[19]

J. C. Cooper, a commentator on Taoism, expands upon the meaning of yin:

> The *yin* principle is the negative, dark side and also symbolizes

> the feminine element, which is the potential, the existential, the natural. It is the primordial chaos of darkness from which the phenomenal world emerged. . . . The *yin* is the eternally creative, feminine, the Great Mother. . . .[20]

The creative principle referred to in the first excerpt is *ch'ien* or yang, the male principle. This principle is active and is identified with heaven (the female principle is identified with the earth). It gives form to ideas and is active in all respects:

> Vast indeed is the sublime Creative Principle, the Source of All, co-extensive with the heavens! It causes the clouds to come forth, the rain to bestow its bounty and all objects to flow into their respective forms. . . . Whatever it undertakes, the Creative Principle invariably carries to a successful conclusion.[21]

Intelligence, rationality, motion: all are denoted by the masculine, by the ch'ien, or yang. A psychological interpretation of the system would be that anthropomorphic sex roles and sexual imagery were being used as the basis for the metaphysical system. But to take it in terms of the system itself, this is decidedly not the case. The specifically human male and female are but instances of a more general organizing principle of the universe, that of interdependence between the active and passive principles. Their human manifestations happen to be male and female merely because everything in the universe is controlled by these principles, and therefore it would be natural to expect their manifestations in human life as well. Moreover, there is not a strict sexual division between yin and yang. Both men and women have both principles, although they are dominated by the one that is in accord with their biological sex. Successful living requires a balance of yin and yang energies within each person. Taoism teaches that sexual arousal will stimulate the yin essence in women and the yang essence in men and that orgasm during intercourse will release these energies so that they can be absorbed by the partner through the sex organs. Various sexual practices arose to maximize the possibilities of this exchange.

Although balance is the aim of Taoist practices, the mystical conceptions of activity and passivity (masculinity and femininity) can be seen as a divisive influence in the conception of sex roles. If we look at the system from a functional perspective, we can see that these symbolic ideas reflect the sex role system, especially with regard to reproductive roles (the image of female as maternal figure is particularly strong). One could further argue that these conceptions serve to justify and maintain a social system based on sex role division. The Taoist answer would doubtless be that even though all of the above might be true, the most important aspect is the fact that the yin and yang principles represent a mystical insight into the nature of reality, a true inference about the working of the universe, which in its human aspect reflects yin and yang in the sex role division of male and female. Since the ultimate nature of reality is unknowable (at least as approached through the consensually validated proofs of rationality and observability), we have no generally agreed-upon way of ascertaining the truth of the more general Taoist vision. In any case, even if it is "true," the truth of the psychological and social link to the sex role system cannot be disputed. Any pervasive conceptual system must have an influence on the way men and women think about themselves and the other sex, especially if the differences are stressed through sexual practices. In turn, if one concedes that the mystical insight into reality was expressed by mortal men (probably not women), it is only reasonable to suppose that the form of these insights was determined by what they experienced in the real world of sense impressions. What this meant as it concerned sexuality was strict sex role separation with a modicum of interdependence, especially with regard to sexual satisfaction and reproduction.

The Taoist case is not idiosyncratic, however. Throughout the world we see instances of complementary systems of male and female principles. Eliade describes the "cosmic hierogamy," the marriage between the (male) Heaven and the (female) Earth, as a common theme found in Oceania, Asia, Africa, and the two Americas. Other archaic religions put forth the vision of an androgynous divine entity. According to Eliade:

> . . . the phenomenon of divince androgyny is very complex: it signifies more than the co-existence – or rather coalescence – of the sexes in the divine being. Androgyny is an archaic and universal formula for the expression of *wholeness*, the co-existence of the contraries, or *coincidentia oppositorum*. More than a state of sexual completeness and autarchy, androgyny symbolises the perfection of a primordial, non-conditioned state."[22]

The symbolism of male and female as an interdependent system seems to be pervasive. We can perhaps understand this on a social level if we note that the primary division of labor in all societies is sexual, as is the primary division of reproductive function and physical appearance. One can see these bipolar attributes as the basis for speculation about the nature of the universe, especially in relation to creation (perhaps the primary metaphysical problem).

One interesting twist to the sexual symbolism we have been considering is the Hindu Tantric view of maleness and femaleness. The male here is described as "the undifferentiating absolute to be awakened by feminine energy." The female is the active, creative principle and maternal imagery is the dominant theme. Sex at the human level symbolizes the creative interplay between male and female principles:

> The act of continuous creation is expressed by patterns in sexual activity, which is seen as infused with a sense of totally transcendent love. The existence of the world is thought of as a continuous giving birth by the yoni (vulva) of the female principle resulting from a continuous infusion of the seed of the male, in sexual delight. The yoni is that rocket or monstermouth spewing out the world; but at the same time there would be neither world nor yoni without the seed, which gives the whole system its possibility of existence, its Being, which is always implicit, but can never be an object of perception.[23]

The Hindu Tantric system is incredibly complex and involves numerous personifications of the principles. For example, Brahman is the ultimate matter and is both male and female, but appears as the personification Shiva to humans. Shiva then unfolds into male and female principles. The female principle is known as Shakti, described as "power, ability, capacity, faculty, energy . . . the active power of a deity."[24] Shakti also has a destructive aspect, Kali, since creation is premised on prior sacrifice. According to Eliade, the myth of creation by violent death is a common one throughout the world:

> This myth of creation by a violent death transcends, therefore, the mythology of the Earth-Mother. The fundamental idea is that Life can only take birth from another life which is sacrificed. The violent death is creative—in this sense, that the life which is sacrificed manifests itself in a more brilliant form upon another plane of existence.[25]

The creative and destructive are intertwined in the female principle, and it is just this combination that has characterized many manifestations of the maternal image in myth and ritual. According to Erich Neumann, who has made a study of such imagery, the positive aspects of the image derive from the mother-child relationship, but the negative aspects reveal themselves in the theme of the terrible mother goddess:

> Just as world, life, nature, and soul have been experienced as a generative and nourishing, protecting and warming Femininity, so their opposites are also perceived in the image of the Feminine; death and destruction, danger and distress, hunger and nakedness, appear as helplessness in the presence of the Dark and Terrible Mother. . . . Thus the womb of the earth becomes the deadly devouring maw of the underworld, and beside the fecundated womb and the protecting cave of earth and mountain gapes the abyss of hell, and the dark hole of the depths, the devouring womb of the grave and of death, of darkness without light, of nothingness. For this woman who generates life and all living on earth, is the same who takes them back into herself, who pursues her victims and captures them with snare and net.[26]

The identification of the female with the earth has been a persistent theme in mythical thought. Mother goddesses of fertility appear throughout the world, and some have interpreted this as evidence for a prehistoric period of matriarchal dominance. Recent thinking has tended to dispute this theory,[27] given the total absence of female dominance in present cultures and especially in hunting and gathering societies, which presumably resemble the earlier hypothesized matriarchal cultures. An analysis of anthropomorphic figures from Crete (a reputed site of matriarchy) showed that less than two-fifths were female.[28] Other reports indicate that the fatness of female figures was stressed more than their sexual generativity, suggesting to Pomeroy that "perhaps hunger was more of a concern than sexuality."[29] However, there is much cross-cultural evidence of the mythic and ritualistic importance of female fertility symbols and female identification with the earth. Eliade reports the worldwide occurrence of female sexual symbols, such as caves and grottoes, in association with religious practices. In many societies, it is customary for childbirth to occur on the soil and for newborn infants to be placed in contact with it. According to Eliade:

> . . . giving birth and parturition are the microcosmic version of an exemplary action accomplished by the Earth; every human mother is but imitating and repeating that primordial act by which Life appeared in the womb of the Earth. Therefore every mother ought to find herself in direct contact with the Great Genetrix, and let herself be guided by her in the accomplish-

ment of the mystery of the birth of a new life, so as to share in her benefic energies and her maternal protection.[30]

Childbirth is such an extraordinary event that one can see why it might occupy a central place in any symbolic system. In addition, any metaphysical system must solve the problem of the creation of the universe and of the human race. The prominence of maternal imagery has probably been a result of attempts to explain creation in terms of a metaphor of childbirth. The fertility of the soil is of course of prime concern to any human group and would also contribute to these notions. The constellation of mother-creation-fertility lies at the base of most mythic visions and naturally shapes the cultural view of women as well. One could take a more psychodynamic perspective and claim that the emotional salience of the mother-child tie is so great that it shapes religious and symbolic representations in the social structure through a projective mechanism of individual needs. Since all humans have presumably experienced the mother-child tie, this relationship provides a common thread linking similarly fomed myths and rituals throughout the world. This interpretation has a decidedly psychoanalytic flavor and argues for the primacy of psychological factors over social ones. However, by giving the perspective a slight twist, one can make it more social in orientation. Since the mother-child tie is the only visibly biological linkage within the kinship structure (all other ties, marital, father-child, and sibling, are defined more socially than biologically), it might be a natural expectation that this primordial biological-social relationship would lay the basis for all subsequent symbolic systems. The two perspectives are not exclusive, of course, if we assume that even a modicum of emotional content could so charge this mother-child tie that its societal centrality would propel it into prominence both psychologically and sociologically. We need not assume that all cultures have the emphasis on maternal love that ours does to argue this case. Just physical dependence and closeness during feeding could be enough, coming as they do in the early formative years, to give this tie a lifelong salience. In addition, the inevitable intermingling of reward and punishment, closeness and withdrawal, inherent in the tie could account for the ambivalent feelings lying behind the twin symbolic images of good and evil in maternity. As we have seen, these images are actively present in the Hindu personage of Kali, and subjectively so in the Taoist tradition, as well as in many others. To go farther out on a limb, one could speculate that the common symbolic images of androgyny and later separation from the perfect unsplit state could refer to the uterine experience of wholeness with the mother and its subsequent end at birth (or perhaps even the break between childhood and adulthood). Often the period of wholeness is viewed as an original, primordial state of goodness replaced at separation by the appearance of evil. We shall see these images appearing in the Western traditions of Judaism and Christianity.

The Old Testament gives two accounts of creation, the Garden of Eden story familiar to most of us, with its account of Eve being created out of Adam's rib, and an earlier (in sequence, but chronologically later) account of a simultaneous creation: "God created man in His own image; in the image of God he created him; male and female." (Genesis 1:27). Interpretations of this passage have varied from that of actual simultaneous creation of male and female human beings to that of creation of one being, "man," with the bipolarity of male-female being a reflection of the androgynous nature of original man or of the embodiment of God within him/her. The sin of Adam and exile from Paradise can then be seen as signs of separation from the original wholeness and entry into the human world of sin and death, as well as sexuality (since the Bible tells of Adam and Eve becoming aware of their nakedness, and presumably of their sexual impulses, only after the Fall). In Jewish thought, the exile of the Jews and the earlier exile of Adam often become symbolically intermixed. In the Jewish mystical writings of the Kabbalah, the image of God is given both a masculine and a feminine (*Shekhinah*) aspect, originally unbroken in androgyny, but then separated through human sin and exile. Gershom Scholem, the leading contemporary writer on the Kabbalah, comments on this step from unity to separation:

The exile of the *Shekhinah,* or in other words, the separation of the masculine and feminine principles in God, is usually imputed to the destructive action and magical influence of human sin. Adam's sin is perpetually repeated in every other sin. Instead of penetrating the vast unity and totality of the *sefiroth* (aspects of God) in his contemplation, Adam, when faced with the choice, took the easier course of contemplating only the last *sefirah*. . . . Since then there has been, somewhere deep within, a cleavage between the upper and the lower, the masculine and feminine.[31]

The theme of original unity and subsequent separation through creation is common throughout world symbolic systems and can be related to the universal human experience of birth and separation from the mother. The Tao, for example, is conceived of as the original primordial substance, subsequently separated into "masculine" and "feminine" elements (although the Tao itself is described with a preponderance of feminine imagery). In the Jewish Kabbalistic writings, the elaboration of the feminine aspect of God is quite explicit. Described as the tenth emanation or *sefiroth* of God, the Shekhinah is described in female metaphorical terms, at times identified with the wooed bride in the Song of Songs, at other times given other female positions, such as mother and daughter. In its positive aspect, the feminine Shekhinah is described as the dwelling place of the soul, but a negative aspect is present, too: the *Zohar* (part of the Kabbalah) says "At times the *Shekhinah* tastes the other, bitter side, and then her face is dark." Scholem notes here the moon imagery found throughout the world in symbolic depictions of woman.[32] We can also note here the reappearance of the dual good-and-evil sides of the symbolic feminine.

The theme of maternity is perhaps nowhere more obvious in Western religious tradition than in the image of the Virgin and Child, and especially in the medieval cult of the Virgin. From the eleventh century to the end of the middle ages, the image of Virgin and Child became central to the Christian vision.[33] Shrines and chapels were dedicated to her, and as Our Lady her maternal protectiveness was sought for innumerable undertakings. The Virgin and Child imagery became a dominant motif in painting and statuary, and some writers have claimed that the maternal image of mercy and protective kindness was more venerated by the common people than were the members of the Trinity. Before the advent of Christianity, cults of virgin goddesses were common. Virgin goddesses appear in Greek mythology and are more positively viewed than their more sexual counterparts. Sarah Pomeroy, a classical scholar, has speculated:

> A fully realized female tends to engender anxiety in the insecure male. Unable to cope with a multiplicity of powers united in one female, men from antiquity to the present have envisioned women in "either-or" roles. As a corollary of this anxiety, virginal females are considered helpful, while sexually mature women like Hera are destructive and evil.[34]

The common motif uniting virginity and maternal protectiveness may be the child's asexual vision of his mother, and his later denial of her sexuality as a defense against Oedipal feelings. At least this would be the Freudian view. Another interpretation would stress the disruptive aspects of sexuality and its evil connotations in early Christian thinking as reasons for the virginal image of maternity.

The maternal image in Christianity has also tended to become the idealized one for the Christian woman. Catholicism in particular has stressed the ideal image of the Virgin Mary as embodying the maternal characteristics to be emulated by all women. Another, darker, image of woman lurking in Christian thought is that of woman as sexual temptress and weak character subject to the will of men. The Reformation did not bring any dramatic change in this position. Martin Luther reaffirmed the essential Christian view:

> Through the Holy Spirit Adam called his wife by the excellent name of Eve, that is, mother. He does not say woman, but mother, and adds "of all the living." Here you have the true distinction of womanhood, to wit, to be the source of all living human beings.
>
> . . . woman was created for domestic concerns but man for political ones, for wars, and the affairs of the law courts.
>
> Taking a wife is a remedy for fornication.[35]

We can see this symbolization of woman as effect and cause of the social system, which reflected deep sex role divisions. The positive view derives from the maternal role, the central one for women. Such divisions would tend to engender negative views and hostilities, as we have indicated in the preceding section on witchcraft. Unfortunately, some scholars have been so taken in by their own cultural indoctrination that they see the source of such universal images as arising from real experience rather than sex role division and associated projected views. For example, consider G. S. Kirk, the eminent authority on myth, on this issue:

> Women as an expensive and deceitful luxury, suborning men by flaunting their attractions, are a folktale-type motif; for the idea is widespread and traditional, involves no special fantasy, and arises out of everyday experience.[36]

What has happened to masculine imagery? It is central to myth and religion, but because the male is taken as the standard case and the woman as "the Other," to borrow Simone de Beauvoir's term, it is not so obvious to us since we expect to see maleness in a central role. In all Western religions, the Deity is given a masculine image and the human prophets and spokesmen, Moses, Christ, and

Mohammed, are all male. Where women appear in the Bible, they generally do so in a subordinate, and often maternal, role. Men represent the general human condition, women the deviation. Ethical and moral responsibility is given to men and expected of them. Their family role, while not forgotten, is not primary to their symbolic function as it is for women. The symbolic and mythic imagery of maleness most often concerns itself with aggressive, positive action, victory over obstacles, and ethical self-realization. Oedipal themes are often present, along with a marked emphasis on the special conditions of the birth of the hero. Rank has identified the prototype of the hero in a number of Mediterranean and Asiatic myths, and this view has been confirmed by the work of Joseph Campbell,[37] who used a wider sample. Rank characterized the hero as follows:

> The hero is the child of most distinguished parents; usually the son of a king. His origin is preceded by difficulties, such as continence, or prolonged barrenness or secret intercourse of the parents, due to external prohibition or obstacles. During the pregnancy, or antedating the same, there is a prophecy in the form of a dream or oracle, cautioning against his birth, and usually threatening danger to the father, or his representative. As a rule, he is surrendered to the water in a box. He is then saved by animals, or lowly people (shepherds) and is suckled by a female animal, or by a humble woman. After he is grown up, he finds his distinguished parents in a highly versatile fashion, takes his revenge on his father, on the one hand, and is acknowledged on the other, and finally achieves rank and honors.[38]

The Oedipal theme of overtaking the father is present, but it can also be interpreted as a more general celebration of ego mastery or of the achievement of adult ambitions over childhood fantasies. In any case, the image of lone male accomplishment is present, contrasted with the female themes of relatedness to others, succorance, and maternity rather than individual growth. The association of maleness with evil is also much more tenuous than that of the feminine mythic connection to it.

When we come to the realm of male anatomical symbolism, much the same tale can be told. We have already considered the active image of maleness in Taoism, and we also see this in other more specifically phallic representations. The begetting function of the male is emphasized in phallic imagery in Hindu and Jewish writings. Phallic worship is reported from Greece and Scandianvia, with an emphasis on the organ's generative and symbolic powers.[39]

In the latter case, the idealization of maleness was often condensed into the phallic symbolism so that it took on a wider ethical and characterological meaning. The phallus as an aggressive symbol, especially with regard to male homosexuality, is also an apparent (but secondary) usage. Eibl-Eibesfeldt[40] reports on the common appearance of phallic figurines as guards (such as gargoyles) and speculates that the display may have derived from the primate mounting threat, which involves penile erection. One could also say that the identification of maleness with aggression might have been enough to suggest the imagery, with the phallus being a natural symbol for the aggressive quality of maleness.

We shall leave the realm of myth and symbol and move on to that of ritual, specifically concerning ourselves with initiation rites and marriage symbolism.

Initiation rites have received their fair share of psychological and anthropological interest. Although widespread in other cultures, particularly preliterate nonwestern cultures, such rituals have not flourished in modern Western societies. Exceptions to this rule in the general area of puberty rites would be the Jewish ceremony of bar mitzvah for thirteen-year-old boys, marking their assumption of full religious duties and entry into manhood, as well as the less popular bat mitzvah for girls. Various Christian confirmation ceremonies occur in childhood but do not have the full social meaning of initiation rites. Rites of baptism and confirmation share some of the general aspects of initiation rites, however, and do mark religious conversion and entry for Catholics and some Protestant denominations. In the secular realm, the sweet sixteen party for adolescent girls and the debutante debut into society for upper-class eighteen-year-old women bear some of the vestiges of such rites but do not exemplify the full domain of stages and physical markings that characterize full initiation rites of passage into adulthood. For Jews, the ritual of male circumcision in infancy is practiced to denote affiliation with a sacred religious covenant, but since the initiated is an infant, few would claim that he would be cognizant of the change of state, so the situation is not analogous to other circumcision rites done at puberty.

Rites of passage and initiation rites can be considered analogous. As identified by Van Gennep,[41] the rite of passage consists of three stages: symbolic or physical separation from the old status, a transitional period, and the entry into a new status. The term "initiation rite" is somewhat more limited, usually denoting a passage from one life stage to another, typically childhood to adulthood, with the connotation of the acquisition of new knowledge (i.e., being "initiated" into a new status that requires new knowledge). A rite like marriage is clearly a rite of passage from one status to another, but it does not exactly fall within the classic meaning of initiation rite. In any case, our concern here is not with semantic distinctions, but with the relevance of such rites for sex role understanding.

Psychological explanations abound for the rather common occurrence of puberty rites. Typically, such rites involve seclusion from the wider social group, along with other age-mates, some tutelage as to the "mysteries" of adult status in that sex, some physical marking, perhaps circumcision, and reentry into the society as one bearing the new adult status. The aspect of the ritual that makes it of great interest to sex role research is the strict sexual demarcation of such puberty rites. In no known society are males and females initiated together, and, in fact, the rituals seem designed to emphasize the "separate tracking" of adult male and female sex roles. Societies differ, of course, in how much prepubertal sex role division is imposed, but at least at puberty itself the divisions are clear and imposed by tradition and ceremony. In our own society, we might note that the peer culture seems to diverge most sharply by sex in adolescence, with values of romance and domesticity emphasized for females and those of competition and achievement for males. It is at this time that adolescent girls often become preoccupied with physical appearance and attracting a mate to the detriment of their school grades and career goals. Males at this age often register improvements in school performance and an increased seriousness about career plans. Some psychologists have proposed that the greatest divergence in sex typing occurs not in early childhood (where most of the research energy is concentrated) but at the break between childhood and adulthood. Certainly in other cultures, this transition is not left to chance; direct tutelage and rite mark the importance of adult sex role division in the society.

In the case of male puberty rites, some physical marking, commonly circumcision, is used to mark the transition. The prevalence of this practice has sparked some creative psychoanalytic explanations. Bruno Bettelheim, in his book *Symbolic Wounds,*[42] expounds the theory that male circumcision and subincision rites are designed to symbolize identification with the female because of male envy of the female "power" of childbirth. Females often keep the details of childbirth secret and this, along with the obvious "miraculous" quality of the creation of another human being, makes men envious of women in their biological capacities. The wound created through circumcision or subincision bleeds like a menstruating woman and resembles in some general way the female sexual organs. Bettelheim further maintains that male initiation is commonly identified with rebirth, a new birth unaided by women. He gives as an example the custom of *couvade*, in which males mimic the symptoms of pregnancy and childbirth during their wives' experience of these states, as part of the general constellation of cross-sex envy and identification.

Bettelheim does not limit his argument to men, but also maintains that women undergo initiation rites that allow a certain amount of cross-sex identification. He describes rites in which the female organ is extended and manipulated to imitate the experience of men. However, he has considerably more difficulty in explaining the meaning of clitoridectomy and other mutilative rites, which seem to lack positive feedback to the woman. We shall consider this issue shortly and shall see that it seems to have more to do with male concern over regulating female sexuality than with anything the women have to do with themselves.

Continuing with the case of male puberty rites, we come to another psychodynamically motivated theory, which has psychoanalytic overtones in terms of the Oedipal complex. Put forth by John Whiting and his associates in 1958,[43] this theory remarks upon the high correlation between the incidence of polygyny and severe puberty rites for males. In polygynous societies, where one man may have many wives, boys will have little contact with their fathers but much contact with their mothers. Often this latter sort of contact is literal, as close mother-child sleeping arrangements in such societies are common. Whiting argues that such childhood experience will lead to a primary feminine identification, which must be broken at puberty through the imposition of severe initiation rites into manhood. In addition, the Oedipal conflict will be strong in such societies because of the close emotional association between mother and child. The initiation rites serve to prevent incest and open demonstations of rivalry with the father (or with other adult men) by fostering identification with the male. Whiting comments:

> Painful hazing, enforced isolation from women, trials of endurance or manliness, genital operations, and change of residence are effective means for preventing the dangerous manifestation of rivalry and incest.[44]

These rites lead to a strong male sex role identity as a defense against the latent feminine identity. Others have identified the attitudes of machismo, where super-masculine values and contempt for women are stressed, as part of the constellation of masculine protest and overcompensation for the power of the initially strong tie between mother and son. Recently, Parker, Smith, and Ginat[45] have challenged Whiting's theory by hitting at the link of initial feminine identification. They studied a Mormon polygynous society in the Southwest United States and concluded that there was little evidence of such identification in father-absent boys (who did not differ from father-present boys; both had strong masculine identities). One might point out, however, that the Mormon religion promotes a strong male-centered ideology that might have overriden any vestiges of female identity in the sample.

The theme of cross-sex identification that underlies Bettelheim's and Whiting's interpretations is an interesting one. Earlier, we pointed out the prevalence of androgynous conceptions of the deity and of wholeness, and this symbolic conception might relate to a universal human longing to sample both realms of human experience, male and female. The argument seems stronger in the case of male envy of women, possibly because of the early salience of the mother-child bond and the child's desire to emulate the mother. Male puberty rites, in any case, conform to this explanation more than do female rites. However, the end result is not greater closeness with women, but rather greater separation from them, since the function of initiation rites in general is to emphasize the specialness of the disparate domains of adult male and female roles.

A more symbolic interpretation of male initiation rites is provided by Mircea Eliade. According to Eliade, the common feature of many such rites is the initiates' symbolic death and rebirth into a new status. He points out that the rebirth is often expressed in embryological and gynecological terms (cf. Bettelheim). The symbolic death is made more dramatic by the subjection of the subjects to a physical ordeal or mutilation and is generally marked by a separation from the initiates' original age group. The rebirth signifies a victory over the symbolic death, and ultimately over actual death. Thus Eliade explains initiation as a reassurance ritual for the fear of death, a fear that would presumably become more real in adulthood than in childhood. Eliade explains:

> . . . death and initiation become interchangeable. And this, in sum, amounts to saying that concrete death is finally assimilated to a transition rite toward a higher condition. Initiatory death becomes the *sine qua non* for all spiritual regeneration and, finally, for the survival of the soul and even for its immortality. And one of the most important consequences that the rites and

> ideology of initiation had in the history of humanity is that this religious valuation of ritual death finally led to conquest of the fear of *real* death, and to belief in the possibility of a purely spiritual survival for the human being.[46]

In more prosaic terms, Van Gennep's stages of separation, transition, and reintegration follow the lines of Eliade's view.

Male initiation rites are somewhat more prevalent than female rites. Explanations for this disparity usually suggest that the onset of menstruation provides a visible change of state for women, whereas no such definitive borderline between childhood and adulthood exists for men, and therefore one must be socially imposed. One might also speculate that the rites serve as well to transmit the cultural values of the male group, and by extension, those of the social group as a whole. Since men make up the ruling social structure in all societies, the transmission of such values and ideas is a cultural necessity. The bonding of males into a cohesive group may be a more pressing social need than that of women, who tend to be separated into the household structure and lack an extensive network of formalized female associations. One could also argue that the sense of male solidarity furthered by the puberty rites intensifies the hold men have over women by giving them a sense of common interest and superiority over women. Often the content of male initiatory ritual includes the ridicule and devaluing of women as well as the direct teaching of patriarchal values. A rather amusing example is given by Bettelheim.[47] He describes the case of the Chaga men of Africa, who claim at initiation to have gained the power over defecation, that is, they foster the belief among women that adult men do not have to defecate. To maintain this illusion, they hide all evidence from women, who still are subject to this bodily "weakness" and hence are considered inferior to men. The women, however, learn of the hoax at their initiation, but are told not to laugh at the men and to keep up the pretense of the truth of the tale.

Of course, we have in some sense argued ourselves into a circle: the initiation rites exist to maintain male dominance, and male dominance is fostered by the belief system generated by male initiation rites. The circle is only a problem if we are concerned with issues of ultimate cause and effect. For our purposes, a functionalist analysis that emphasizes the interdependence of the parts of a system without yielding priority to one over the other will suffice. Our guess, though, would be that the conditions that led to male dominance, probably a combination of male economic power and aggression acting together with female childbearing, served to maintain the male initiation ritual. Once male dominance was the norm, it could act in an independent capacity to support the self-perpetuating institution of male initiation as a new source of male power in each generation. The role of the separation of mother and son should also be emphasized as a measure that ensured that adult male dominance would develop in adolescent males who had previously been dominated by their mothers.

The case of female initiation rites appears to be more difficult to interpret unless we assume that their primary function is for men rather than women. Such an assumption would not be unreasonable, in that one could suppose that the ruling class (men) could impose such rites on women, and that women could rationalize their state in terms of the societal values imposed by men. Such seems to be the case in the Sudan, where Arab Moslem women undergo the genital mutiliation of infibulation.[48] Such an operation involves the almost complete closing off of the vaginal canal at an early age (sometimes as young as six), as well as the removal of the clitoris and other parts of the female genitals. Girls sometimes die from the effects of these mutilations, and although they have been legally banned for some time, they continue to be performed. At marriage, surgical intervention is often required to permit intercourse and later childbirth. The stated function of the rite is to protect the chastity of the women, and, perhaps more important, by extension the honor of the woman's male lineage (or that of her husband). Such honor seems largely to depend on the purity of women. Hayes, who has studied this society, points out that women are considered to be "inherently oversexed" and therefore in need of protection from their own impulses as well as those of males. Women justify the custom in terms of this protection, as well as a symbolic protection against accusations of sexual misconduct. A woman who has not undergone this operation will be always under suspicion; the only successful way to dispel all doubt is, in effect, to make adultery impossible. In fact, according to Hayes, "uninfibulated women are generally considered to be prostitutes in Sudanese society."[49]

The Sudanese case reflects the male concern with virginity and sexual purity, which is a hallmark of many (though not all) cultures. Of course, in societies without effective birth control, women as well as men are given protection by this concern, as otherwise a large number of unfamilied children would be born bereft of economic support. The family system of male support for female childbearing seems to bring as a corollary a judgment against children whose male parentage (social if not biological) cannot be readily assigned or ascertained. In patrilineal and patriarchal societies where the passage of land, wealth, and power follows the male line, such pinpointing of paternity is critical for the maintenance of orderly male rule. A case might also be made for Susan Brownmiller's[50] argument that marriage (and exclusive sexual rights) arose as a way of defending females against rape. The outcome of such concerns seems to be the institution of female initiation rites involving restriction of female sexuality (such as clitorectomy, which would remove the seat of female sexual feelings, and infibulation). Even in societies that do not impose such physical sanctions on women, greater limitations on the freedom of women are often justified by appeals to the sexual protection argument. The double standard in our society, for example, had its roots in such fears of pregnancy and subsequent family dishonor. The use of chaperones and the imposition of curfews is always justified in terms of protecting women rather than men, though the result is a restriction of female freedom and a narrowing of the female sex role. The Moslem custom of *purdah,* in which women are secluded in the home and prohibited from public appearance without elaborate and complete body and face covering, is the logical end of such beliefs. It is in such societies that the equation of female purity and family honor is strongest and the power of men the greatest. Such customs and belief systems obviously presuppose a strong patriarchal culture. One might also note in passing the common emphasis on virgin goddesses in many cultures (see p. 177) and the strange power they seem to exert over the fantasy lives of men. In the Moslem religion, the figure of Fatima, also called the Virgin and described as Mohammed's daughter, is given reverence—and sometimes also a maternal role.[51] The entanglement of virginity and maternity alluded to earlier (e.g., the Virgin Mary) apparently has psychoanalytic roots that affect the symbolic life of men and the actual life of women.

The marriage ritual is obviously important to an understanding of the relative positioning of women and men in any given society. We have already discussed the common pattern of females being exchanged in marriage rather than males (see p. 117). As we noted, in our society, the symbolization of this exchange occurs when the father of the bride hands her over to the care of the husband, and she changes her name from that of her father's lineage to that of her husband's. In preindustrial times and in nonliterate cultures, marriage was often more a concern of the two male lineages than that of the two young people getting married. Some cultures symbolize the economic nature of the transaction through the imposition of dowries or bridewealth.[52] A dowry is paid by the bride's kin to her husband or to the husband's lineage; it represents in some sense the future inheritance a woman might expect to get from her family and which may be paid in advance to secure a suitable husband (or husband's lineage). In patrilineal societies where women do not normally inherit land, a dowry may represent a substitute for such inheritance. Bridewealth, on the other hand, is a common practice in many African societies and involves payment, usually in cattle, from the husband's lineage to that of his wife's. In this case, the exchange is usually thought to symbolize the earnings lost to the bride's father's lineage through the marriage of the daughter. Mair also describes it as "childprice," as "a payment or a gift made in exchange for a woman's fertility."[53]

Symbols in wedding rites often allude to the assumed virginity of the bride and to hopes for her fertility. The wearing of white and the throwing of rice reflect these values in Western culture. In societies that practice strict sexual segregation, wedding ceremonies often are separated by sex, with a consequent deepening and emphasis on distinctive male and female values. Mason[54] describes the Moslem Libyan wedding rite, in which such features are quite pronounced. According to Mason, the males and females, in separate dancing ceremonies, mock the dancing form of the other sex. The men celebrate in a large, central room, the women in a dark, secluded one that affords a glimpse of the male festivities through a covered window or doorway. At only one point is the strict separation of the sexes violated. A female dancer in symbolic virginial white comes into the male room, symbolically representing female sensuality. The males may sing and dance in sexual counterpoint, but only to a certain limit so that female modesty and purity are maintained. A most important part of the ritual is the groom's testing of the bride's virginity, proved by the display of the stained bedlinen to the mother of the groom. Considerable dishonor afflicts a woman unable to produce such proof.

The ease of breaking marital ties often reflects the relative ranking of the sexes. In Moslem tradition, for example, divorce is a male prerogative and does not require the consent of the female; but a woman may not divorce her husband without his consent. In Judaism as well, divorce was a power given only to men, although in exceptional circumstances (e.g., of nonsupport) a woman could petition for the right. In our own culture, a divorced woman is somewhat more socially stigmatized than a divorced man; and in all cases a female is more identified with her marital status than a male (the obvious example is that female titles—Miss, Mrs.—denote a marital status, whereas the singular male Mr. does not; the neutral term Ms. was created to avoid this labeling process).

In our consideration of myth and ritual, then, we have seen several themes emerge. Female evil, sexuality, and maternity have been emphasized as have male assertion and aggression. The problem of the reconciliation of male and female opposites has been elevated to a mythic and ritual plane, reflecting the more prosaic divisions in the world below. Throughout the biological, psychological, social, and symbolic realms, we see a continuity of themes of masculinity and femininity: power relationships, maternal ties, sexuality, aggression, the family. The same ideas seem to arise in many realms to define the separate fates of men and women.

We turn now to more specific examples of symbolic usage: language, literature and the arts, and will emphasize the findings from our own culture.

LANGUAGE

The domain of language—with regard to sex roles—can be divided into two parts. The first concerns the structure of the language itself, while the second involves the different ways men and women use a given language. In the first case, we have examples of the so-called "women's and men's languages" and the allied phenomenon of linguistic gender.[55] Early anthropological and linguistic studies of a few cultures, notably that of the Carib Indians, claimed the existence of separate male and female languages within a given society. Further investigation determined that in all languages that have male and female variants, the differences are rather superficial and never reflect any structural difference in syntactic patterning (such as word order). What do differ are features of pronunciation and the use of various prefixes and suffixes added to words that identify the sex of the speaker and/or the person spoken about. At times these superficial differences make the sound of male and female speech quite distinctive, but the deep structure of the language remains the same; therefore it is incorrect to speak of "different" languages. The forms of words may vary but not the language itself. Japanese is an example of a widely spoken language where such sex-marked features abound. Interestingly enough, when language variants are discussed, the male form is always taken to represent "the" language, while the female form is identified as "women's language." Another case of woman as "Other," even though, of course, women make up half or more than half of any language community.

A related phenomenon that may be more familiar to nonlinguists is that of linguistic gender. French, German, Italian, and Spanish, to name a few of the more familiar cases, all exhibit gender. What this means is that all nouns are given a sex (or possibly a neutral status) and that other forms (adjectives) must agree in gender with the noun. The gender of nouns is often seemingly independent of their meaning. For example, the German word for nose (die Nase) is feminine, while the word for mouth (der Mund) is masculine. Other features of gender in European languages include marking the sex of the speaker and the person spoken about. For example, a female describing herself as rich in Spanish would use "rica," while a male would use "rico." When speaking of a group of people as "they," English does not specify sex, but other languages do. For example, in French "elles" is "they" for a group of women, while "ils" refers to a group of men. Interestingly enough, in the latter case, if there is a mixed-sex group, the male form "ils" must be used, never "elles," a term reserved exclusively for women or objects of feminine gender.

Some linguists have considered male and female form variants and gender to be related phenomena. Frazer, for example, in writing in 1900, speculated that gender might be the modern remnant of earlier more sex-differentiated language forms. Bodine, however, writing in 1975, argues that they may be "different manifestations of similar social, psychological, or cognitive tendencies"[56] and not historically related. She does not hint at what these tendencies might be, however. The widespread observance of sexual distinctions in speech may simply reflect the overriding salience of the sexual dimension in thought and social structure. Language presumably develops out of the structure of the human mind and of psychological and social experience, and much of the latter two domains is demarcated by sex. Since language arose to fulfill human needs for communication in social situations, if those social situations were sharply divided by sex (and the society organized by it) the resultant language was bound to be as well. Societies characterized by a low degree of social differentiation by any dimension other than sex would develop languages reflecting sexual distinctions. In European languages that make the "tu-vous" distinction between the familiar and polite form of address, we see the dimension of social distance (i.e., status) and solidarity forming the language. It would be natural to suppose that sex would have a similar influence. Once developed in a culture, language forms sharply demarcated by sex could be expected to help maintain the structure, though it is likely that linguistic change follows rather than precedes structural change in the society.

Another aspect of the form of the language is the use of sex in "marking" certain words. In English, the masculine is the unmarked case, so that the use of certain terms ostensibly bisexual in reference (such as poet, chairman, etc.) implies that the individual described is a male. When a female is specifically denoted, an identifying suffix must be added to "mark" the case (e.g., poetess, chairwoman). Related phenomena include the extraneous marking of words otherwise thought to apply to men as the universal case, e.g., "lady doctor" or "woman lawyer," since the unmarked word would presumably refer to a male. A more obvious example of this tendency to identify the universal case with maleness is the use of "he" to denote any person of unknown gender as well as the general case covering all instances. For example, we say "Each person shall hang up his own coat" even when we refer to a mixed group of men and women. Mary Ritchie Key quotes a rather humorous instance of this tendency, taken from a book for children:

The Secret

We have a secret, just we three
The robin, and I, and the sweet cherry tree;

But of course the robin knows it best
Because HE built the – I shan't tell the rest
And laid the four little – something in it –
I'm afraid I shall tell it every minute[57]

Humor aside, we can see the effect such linguistic forms might have on women and men. The use of the universal "he" renders women invisible in the universal case, thus reinforcing their image as "other." Both men and women, under the subtle and insidious influence of language spoken and read day after day, are shaped to see the world through masculine eyes and to regard the feminine as the deviation rather than the norm (or at least part of the norm).

The second aspect of language we shall examine is that of the differential usage of language by men and women, aside from questions of form inherent in the language (such as gender or marked and unmarked cases). Here we shall present English examples for the convenience of the reader. A growing number of feminist linguists have identified specific features of language usage by women and men with sex role distinctions in the social structure. Perhaps the best-known commentator has been Robin Lakoff, whose book *Language and Woman's Place*[58] characterizes female speech as much more polite and less assertive than male speech. She maintains that "women's speech is devised to prevent the expression of strong statements."[59] Lakoff argues that women use more politeness markers, such as "Please," "Would you mind. . . ," etc., as well as stating opinions in the form of questions, such as "The economic situation is bad, *isn't it*?" That latter case includes a tag question, where the statement of opinion is softened by the appended inquiry and implicit appeal for agreement. Men, Lakoff maintains, are much more likely to be direct in their requests and to state opinions flatly, indicating their dominant social position.

Other research seems to support the view that women may be

more oriented to the needs of others in the social situation while men seem more self-directed and autonomous of it. In a study of same-sex groups, Aries[60] found that female groups often focused on interpersonal concerns involving feelings and personal matters, while male groups spent most of their time on nonintimate and interpersonal topics, favoring themes of aggression and competition. In mixed-sex groups, Aries found that "the presence of women changes the all male style of interacting, causing males to develop a more personal orientation, with increased one-to-one interaction, greater self-revelation, and a decrease in the aggressive, competitive aspects of the encounter."[61] In my own study of sex differences in nonverbal communication,[62] I found evidence for a more partner-focused style for women, with the possibility that women may adjust their nonverbal style of communicating to fit the needs of men.

In general, then, the feminist perspective on language emphasizes the more constrained style of female speech as symptomatic of women's oppressed state in society. It should be emphasized that little empirical work has actually been done to date to back up these observations on sex differences. One study[63] of conversation topics introduced by men and women in couples found that the topics introduced by the women tended to be dropped in favor of those introduced by men. That is, men were more successful in getting the couple to discuss what they had in mind than women were. The theme of adjustment and orientation to others rather than assertion and self-starting seems to characterize thought on female speech styles. Of course, these styles are only linguistic in a superficial sense; the real variables under study are motivational and interpersonal. For example, if women defer to men in conversation topics, this could mean both that men refuse to be deflected in their conversational goals and that women devalue their own status and find pleasing the male more important than getting their own point across.

In an extended study of cross-sex speech interactions, Zimmerman and West[64] found that men tended to interrupt their partners more than females and to use such interruptions as means to control conversational topics. In addition, the pattern of silences in cross-sex interactions tended to be asymmetrical: women were silent more than men. The researchers assert: "men deny equal status to women as conversational partners . . . just as male dominance is exhibited through male control of macro-institutions in society, it is also exhibited through control of at least a part of one micro-institution (conversation)."[65]

Where do such linguistic sex differences come from? The most obvious source is the structure of sex roles themselves. Women are defined more tightly in society; they are more constrained in how they may act; therefore they must be more polite and accommodating. The commonly used term "lady" denotes these qualities of social acceptability and social control much more so than does the equivalent term for men, "gentleman." The subordinate role is that which must curry favor with the superior through the use of courtesy and constraint. We have many examples of asymmetrical role relationships between men and women, both on the individual and the societal level. One linguistic marker of this constellation is the (unfortunately) common usage of the term "girl" to denote a female of almost any age, whereas "boy" is reserved for young male children. In fact, the use of "boy" to address black men was considered to be an especially degrading appellation. A similar instance is the familiar use of first names, a more common occurrence for women than men. In these little linguistic signs we see the structure of a sex role system emerging.

An alternative and somewhat complementary theory of linguistic acquisition is the modeling hypothesis. Language can be learned like any other behavior and can be imitated from adult model usage. Some evidence for this idea comes from Jacqueline Sachs's research,[66] in which she found that girls' and boys' speech could be readily distinguished by adults even though the anatomy of male and female vocal tracts (before puberty) seems to be identical. When boys and girls four to fourteen years old recorded the same sentence, 81 percent of the adult guesses as to sex were correct. In analyzing the tapes, Sachs and her colleagues found that the formant values for some vowels were significantly lower for boys than girls. This pattern is also found in adult male and female speakers. In the absence of obvious anatomical sources for the difference, the researchers speculate that the children could be mimicking adult voice models. For example, Sachs points out that the female tendency to smile while talking could produce the formant pattern observed. (One might note in passing the overall greater frequency of female smiling as a cultural characteristic and as a possible by-product of an accommodative interaction style for women.)

We turn now to a consideration of literature and the arts in relation to the sex role system.

LITERATURE AND THE ARTS

The cultural products of any given society at any given time reverberate with the themes of that society and that era. One theme that underlies much artistic creation is that of sexuality and the social roles of men and women. We can only skim the surface of this rich area, which is only now getting the attention it deserves.

Feminist criticism of literature has aimed at the stereotypic notions of femininity found in many of the writings of well-known authors. For example, D. H. Lawrence has been viewed as having an especially bifurcated view of femininity and masculinity, identifying the first with emotional richness and closeness to nature and the second with instrumental power and phallic assertion but lack of connection to the earth. He views sexuality as the necessary merging of opposites, but with an emphasis on submission in women that has angered feminist critics. Such images of masculinity and femininity recall the mythic images of opposition and maternity discussed earlier.

In more direct terms, novelists have portrayed women and men in their societal roles, which reflect different domains and different character traits. Few literary images of independent women, defined apart from men, exist; they are far outnumbered by domestic visions of womanhood. Yet novelists, particularly female novelists, have recognized the problems of marriage for women. Patricia Spacks writes of the nineteenth-century English novel:

> Marriage, for a woman, means dependency: the appeal of irresponsibility, or of responsibility defined by others, and of irresponsible power; the sacrifice of autonomy; endangered inner experience. Women novelists recognize the appeal, but also the threat. In their investigations of internal and external female experience, they often question, overtly and covertly, marriage as a happy ending—even in the context of Victorian propriety.[67]

As Spacks terms it, "power through passivity," exercised through marriage and through men, became the overriding concern of many female writers. The male as hero and character in literature is seen as more autonomous but also subject to the matrix of social determinants. Literature is life; literary plots and characters reflect societal situations and images. The creators of such literature see the world in perhaps a more exacting or incisive way than most, but what they see and write about strikes some accord with reality or else readers would not respond. Some writers, August Strindberg, for example, may have a particularly pathological view of women, but that view must reflect some cultural current of thought if it is to have any meaning to readers. Literature does not create society or sex roles; it merely codifies what already exists and highlights themes of meaning to writer and reader alike.

Great literature has always had a rather small audience. If we are concerned with the social impact of literature, we had best turn to writing that reaches the mass market. In recent times, magazine fiction has accounted for a large proportion of fiction readership. Women make up the majority of magazine readers and have been fed a steady diet of fiction glorifying the domestic role, at least for the last thirty years or so. In her book *The Feminine Mystique,* Betty Friedan identifies the common theme of women's magazine fiction: "Fulfilment as a woman had only one definition for American women after 1949 – the housewife-mother."[68] In an analysis of stories appearing from 1940 to 1970, Franzwa[69] found several significant themes emerging. When the story concerned a young single woman, she was almost always portrayed as concerned about marriage, as less competent than men, and as patient and passive. Work as a goal was not stressed for unmarried or married women. In fact, in the case of married women, work was seen as a negative force, a threat to the male ego and the structure of the family. Having and raising children and managing a household were defined as the only worthwhile goals for (married) female life. To be other than (happily) married was an unthinkable fate for magazine fiction heroines. Children's books also present this image of womanhood and the contrasting achieving image for men: "Boys invent things. Girls use what boys invent," states one book for children.

Popular books and fiction aimed at men tend to present the male as unfettered by family ties, going off to some solitary adventure (perhaps accompanied by an attractive female companion, definitely not a wife). Western novels by Zane Grey and others promote this image, as does detective fiction, perhaps the most glamorous to date being the James Bond series by Ian Fleming. Few fictional portrayals of men concern themselves with family ties, although almost all of women do. One exception to this rule for men appears in some television series (such as the old "Father Knows Best") where family responsibilities are stressed (with the male as uncontested head of the family), but these are counter-balanced by the solitary-male-hero-type series that predominates. Aggression is perhaps the dominant theme in male popular fiction. War exploits, adventure tales, westerns, and detective mysteries all feature violence (usually in the service of some moral good) in a central role. Women, if present at all in male fiction, tend to be cast in stereotyped sexual images. Female fiction, on the other hand, tends to feature men in prominent positions, and the happiness of female characters is almost always in some way dependent on male actions.

The visual arts and media are also an arena for the playing out of societal sex role themes. In European painting, religious motifs were dominant for many centuries. The Christian themes of Virgin and Child and of Christ and the Disciples are noteworthy for their emphasis on the maternal but secondary image of women as opposed to the primary masculine ideal of God, Christ, and the Disciples. We have already discussed these themes in an earlier section, and since we view art as a social product, we add little more here.

The portrayal of the erotic in art is a theme rich in sex role connotations. In Western art, the female nude has appeared frequently, with an emphasis on her sexual characteristics and appeal to a male audience. The male perspective was primary: "Controlling both sex and art, he and his fantasies conditioned the world of erotic imagination as well."[70] Various erotic images of woman compete for our attention: the maternal warmth of Renoir's nudes, the femme fatale image of Munch and of the "Gibson girl," with its intimation of danger in female sexuality. Above all, the ideal of female passivity and availability for male sexual pleasure is expressed. As Berger puts it, in describing the erotic effect of a painting of a female nude:

> The painting's sexuality is manifest not in what it shows, but in the owner-spectator's . . . right to see her naked. Her nakedness is not a function of her sexuality but of the sexuality of those who have access to the picture. In the majority of European nudes there is a close parallel with the passivity which is endemic to prostitution.[71]

Of course, male nudes have also appeared in art, notably in the sculpture of ancient Greece. In those cases, there was an attempt to equate ideal male physique with the ideal of humanity and of human proportion. Some have also speculated that the homosexual ethos of ancient Athens fostered the interest in the nude male figure. It is almost certain that appeal to women was not an aim of the work, as the audience and artists were defined as almost strictly male.

In the visual media of today, film and television both reflect and shape our views of sex roles. Both media attract a mass audience; the personalities and situations portrayed become part of popular culture in short order. Films and television can transmit ideas and images to millions of people at once; their symbolic language becomes our language, their ideas and ideals, ours. Because much of the content of both media has great relevance for sex roles, film and television assume prime positions of importance in our consideration of influences on the sex role system.

In the realm of film, much recent criticism has focused on the view of women portrayed. In the early days of film, a romanticized, unreachable woman was featured, to be succeeded by more sexual characterizations. Rarely were sexuality and any intellectual dimension of character combined in female characters. An exception emerged in the 1940s when favorable images of successfully achieving women emerged. The common appearance of women in jobs during World War II has been credited with this change. The change was rather short-lived, however, and after the war the familiar dichotomous images of bad woman–good woman appeared. The bad woman was alluring and sexually provocative, but empty-headed and basically immoral. The good woman was domestic and oriented to husband, home, and children, but sexless. In the 1950s and 1960s, Marilyn Monroe and Doris Day exemplified these images. Of Monroe, Molly Haskell has written:

> She was partly a hypothesis, a pinup fantasy of the other woman as she might be drawn in the marital cartoon fantasies of Maggie and Jiggs, or Blondie and Dagwood, and thus an outgrowth, once again, of misogamy. She was the woman that every wife fears seeing with her husband in a convertible . . . or even in conversation, and that every emasculated or superfluous husband would like to think his wife lives in constant fear of. She was the masturbatory fantasy that gave satisfaction and demanded nothing in return; the wolfbait, the eye-stopper that men exchanged glances over. . . .[72]

Many less talented actresses followed in Monroe's footsteps and made the sexpot image part of the currency of American life and of American womanhood. The view of women suggested by such an image is that of the totally sexual–no other intellectual or emotional dimension could interfere. Ironically, though, the sexual image was not one of female sexuality, but of male sexuality served. As Haskell put it, "she was the fifties' fiction, the lie that a woman has no sexual needs, that she is there to cater to, or enhance, a man's needs."[73] The sexpot had no self; her definition was in the eyes of the male beholder. Besides the films themselves, their stars, such as Marilyn Monroe, came to stand for this image of women. As promulgated through film, fan magazines, newspapers, and other media, the sexpot image became one of the most basic dimensions of the symbolic female sex role, and one that every woman had in some way to relate to. The American preoccupation with breast size, as shown by the prevalance of "training bras" for twelve-year-olds and the concern shown for a woman's "vital statistics," is an example of the identification of the female with her body. The attitudes that led to such an image were not created by the movies, but their worldwide influence in shaping (literally) our view of women and glorifying it into a national ideal cannot be denied.

Coexisting with the sexpot image was that of the domestic woman, perhaps best exemplifed by the roles of Doris Day. Such a woman was the perfect embodiment of the feminine mystique, seeking love and marriage and motherhood as her defining qualities, with the central presence of a male needed to bring meaning to her life. Like the sexpot, this image was processed through male eyes, since it was to their welfare that such a woman was dedicated. However, the domestic image perhaps did represent to a large degree the American woman's image of herself, since the way had been prepared by earlier purveyors of feminine mystique values, such as advertising and women's magazines. Many women were in fact living out the values of such a domestic lifestyle, although one suspects they did not all find the complete satisfaction afforded the likes of Doris Day-type characters. Such portrayals were not sexual nor were they intellectual or assertive. Pliant and pleasing, dainty and domestic—such were these characterizations.

In the late 1960s and the 1970s, a new image of women emerged. The Women's Liberation movement may have helped tear asunder the domestic values of the fifties, but it was not successful in replacing them with a cinematic image of a more complete woman. Instead, as Joan Mellon points out, the new women as portrayed in film become sexually active and independent of the constraints of family life, but the values of the family are reasserted in that such women are negatively portrayed and are often punished. Mellon comments:

And as the sexually compliant and uninhibited women are clearly amoral, their punishment in sadistic violence takes on a quality of vengeance on behalf of the home and the hearth.[74]

Mellon believes that the negative image of woman was always present in cinematic portrayals, even in the domestic portrayal:

> The façade of protection has been withdrawn and the real, unconscious feelings toward women emerge. All along, it seems, domesticated women were resented. They were resented by men because they were sexually disappointing and also because their portential sexuality was both desired and feared.[75]

Mellon's view, like those of many other feminist critics, is that men have dominated the film industry and thus determined the image of woman presented. To a large degree this is true, as women producers, directors and writers have played almost no role in Hollywood. Also true is the fact that the audience aimed at in the portrayal of women is men. Men largely determine what portrayals of women will be successful. The sexpot image is clearly a case in point. In the days before Women's Liberation consciousness (and even after it) many women also accepted this image, even though it was not of their own making. The image did feature orientation to male needs, always a major part of the female sex role in modern America, so it probably struck a responsive chord. The rest of the culture did not present any counterimages of whole, independent, functioning women, so the negative cinematic images would take over the field of male (and female) imagination.

Film images of masculinity of course abound, although, interestingly, there is usually little separate critical treatment of them (as contrasted with that of women). Several genres of film, for example the western, the gangster movie, and the war movie, tell us what masculine qualities are given central prominence. The aggressive hero emerges as the prototype of manhood. Such aggression is usually justified in the service of some greater moral good, such as the use of legitimate police power against illegitimate gangster power, but the violent motif reigns supreme and often little difference can be found between the two sides. In the case of the western, the themes of white man against Indian and white man against lawlessness predominate. War movies retell historical events with an order of plot not found in real-life battles, so that heroism and dedication clearly emerge. In all three of these genres, western, gangster movie, and war movie, the fellowship of men with each other is primary. In all cases men are seen working together, often in organized groups, gangs, posses, platoons. Their primary reference group is each other; women intrude only to introduce a momentary pause for sex or domesticity. Family ties are generally viewed unfavorably in these films and are seen as a distraction from the real work of men. In the western in fact, the West is often symbolically represented as a masculine escape from the feminine East, a free place where men can be men without women (or at least without the enduring ties to women that family life entails).

Men are allowed to exhibit qualities other than violence, although the threat of force is usually viewed as the ultimate arbiter of right. In detective and mystery movies, the cunning intelligence of the hero is pitted against the evil (but usually intelligently portrayed) machinations of the criminal. In almost all cases, both the seekers and the sought are men. Women appear as victims of crime and as ancillary figures to the primary male actors.

The image of the male as lover has of course appeared on the screen as well, from the days of Rudolf Valentino to those of Paul Newman. Rarely has the male been defined totally in terms of his ties to women, however; usually the love theme is a subsidiary one to the main concern of the man. When women appear, however, they are usually defined largely in terms of their love and domestic lives and their roles as sweetheart, wife, or mother. Their reference group is never other women; rarely do several women appear in a film concerned with a common endeavor, as is often the case for men. When women do appear, they are featured for their sexual or domestic function. Male characters are usually developed to a far greater degree because of their central role in the film.

Cinematic images of masculinity, like those of femininity, both reflect and support the existing sex role system. Themes of aggressive achievement of some moral goal and of victory over obstacles through cooperative interdependence with other men predominate. Family life and relationships with women are given secondary status, defining the way the ideal types of sex roles are viewed in this society. Girls and boys growing up in such a symbolic atmosphere see the reaffirmation of socialized sex role ideals on the screen of their neighborhood movie house. The world of the movies is a simplified world where the complexities of life are sorted into simply plotted packages. Within these packages lies a definite message about sexually divided roles. The realms of sexual allure and domesticity belong to women; those of aggression, moral achievement, and intelligence, to men. In addition to the films themselves, the more successful actresses and actors come to personify the traits they so often portray, and as idols of the masses wield additional influence on their audience. Actors like John Wayne, for example, become symbols of certain sex-typed aspects of masculinity that are valued in the culture.

Even more pervasive than the influence of the movies is that of television, since exposure to it is so much greater, and in a sense, more personal, given the home setting. Television characters are met week after week in the same roles in the same series and take on a familiarity not matched by any other medium (with the possible exception of radio in the early days, but the absence of video probably lessened the impact somewhat). Many of the film genres find their match in television, particularly the western, crime, and adventure dramas, with similar messages about sex roles. Several new forms have emerged on television (based to some degree on earlier radio formats): the situation comedy and the soap opera. The situation comedy presents a world with strict sex role lines. In the 1950s the prototypical show was "Father Knows Best," with other programs, such as "Ozzie and Harriet" and "The Donna Reed Show," presenting similar views. Horace Newcomb has described these roles:

> The father stands at the center of the family. In the most traditional sense he provides leadership and wisdom, and it is not so important that Father knows *best* as it is that Father *knows.* Much of his wisdom and authority rise from the fact that he has a much stronger definition of his function outside the home.
>
> [The mother] reassures us by acting as the provider of physical comforts within the home. Though she is not the primary judge of actions within the family, she is often the source of behind-the-scenes wisdom. . . . Her wisdom is increased by her choice to allow the father the appearance of superiority in his role, even though she has directed the decision.[76]

In more recent times, with the emergence of "modern" situation comedies like "All in the Family" and "Maude," sex role divisions remain secure. In "All in the Family," for example, the household revolves around Archie, who is presented in a caricatured role but whose views must be taken into account by the domestic Edith. It is she who accommodates his craziness, not the other way around. In the case of the younger couple, Mike and Gloria, the sex roles of student-achieving-male and supportive-domestic-female emerge with great clarity. Even in the innovative "Maude" show, the power of the central (female) character is limited by the indulgence of her husband and the implicit constraints of the nuclear family system.

The soap operas follow somewhat the same scenarios as the situation comedies, but without the comedy and with more serious and involved plots. Problems do not get solved within the time limits of the half-hour time slot as they do in the situation comedy. The problems of life presented, those of marriage and family and death, do not lend themselves to easy answers, and the agonies of the characters are stretched out over the years of the programs' survival. The roles presented in such serials are sexually demarcated although women are often portrayed in a more sympathetic light than in the situation comedies, perhaps because they are afflicted with such weighty problems (and also because almost all the viewers are female). Many of the problems of women still involve relationships with men, and tales of illegitimate births and extramarital affairs abound. The audience for such fare is limited almost entirely to women, in contrast to other television shows, and many women develop an addiction to watching and worrying about soap opera characters that defies description. The image of the housewife worrying about the fates of soap opera characters (fates far worse than hers) is almost a cultural stereotype.

Children's television is a particular area of concern for sex roles, although of course much of the nominally "adult" fare is viewed by children as well. Studies have shown that few female characters appear in other than stereotyped roles on children's television, even in supposedly "liberated" programs such as "Sesame Street." Male characters abound and reflect some of the achievement through violence found in adult shows. In fact, most of the controversy over children and television has concerned the prevalence of aggressive content in many shows and its possible pernicious effect on the young. Needless to say, such aggression is almost always male aggression, found within the genres of the western, crime, and adventure dramas. The sex role messages children are getting from television are very strong and are given over a long period of time of relatively constant exposure. The average child spends almost as much time in front of the television screen as in school and will often give more credence to the message of the former than to that of the latter. The seductive power of television is hard to deny; it has become a force in the national life as well as in the individual socialization experience of the child. The images of the television screen lurk in the symbolic consciousness of the child and the adult and reflect and reinforce the sex role reality of the world outside.

Television commercials and advertising in general contribute importantly to the maintenance of sex roles. The largest number of products portrayed are aimed at women, the primary consumers. Household cleaners and headache remedies abound. Male announcers often speak with authority to bewildered women who cannot tell the difference between two brands of cleansers or aspirin. The implicit messages of male authority and female dependence and domestic orientation are inescapable. Advertising aimed at men often promotes the image of sexual conquest and aggressive competitiveness. Thus, car ads often depict comely females draped over sports cars, and professional athletes are brought in to extol the virtues of hair tonics. The well-known image of the "Marlboro Man," with its suggestion of frontier masculinity, has sold innumerable cartons of cigarettes. The advertising folk are a pragmatic sort who know that sex-role-tinged messages work: they sell products. They work because of all the factors we have discussed so far; the society is geared to a traditional notion of sex roles, and it is these images that have been largely internalized and accepted by the American public. They see themselves the way the culture sees them.

In this chapter we have explored the symbolic world of sex roles as shown through taboo, witchcraft, myth, and ritual, and more immediately through language, literature, and the arts. Throughout we have seen similar themes emerge: a concern with the sexuality of women and maternity, with the reconciliation of sexual and societal opposites, with male aggression and power. The symbolic vision of sex roles reflects the worldly one and helps maintain the integrity of the sex role system through imposing on the psychological plane those divisions that exist in the social structure.

Now we move on to our final chapter, which considers sex role change through space and time. We have seen how the biological, psychological, and social systems work together to maintain the structure of sex roles. Certain societies at given points in time have tried to introduce change into this structure. The nature and fate of these changes, seen both from the cross-cultural and the historical perspective, shall form the substance of the next, and final, chapter.

SUGGESTED READINGS

Bruno Bettelheim. *Symbolic Wounds: Puberty Rites and the Envious Male.* New York: Macmillan Co., Collier Books, 1962 (p). Provocative psychoanalytic theory including some anthropology.

Paula Brown and Georgeda Buchbinder, eds. *Man and Woman in the New Guinea Highlands.* Washington, D.C.: American Anthropological Association, 1976 (p). Collection of five studies on an area of the world rich in sex role symbolism.

Vern L. Bullough. *The Subordinate Sex: A History of Attitudes Toward Women.* Baltimore: Penguin Books, 1973 (p). Interesting survey of women's place in symbol and in religion; historical material as well.

Mary Douglas. *Purity and Danger: An Analysis of Concepts of Pollution and Taboo.* New Orleans: Pelican Books, 1966 (p). Much here that is relevant to sex roles.

Mary Ellmann. *Thinking about Women.* New York: Harcourt Brace Jovanovich, 1968 (p). Women as portrayed in literature.

Claire R. Farrer, ed. *Women and Folklore.* Austin, Tex.: University of Texas Press, 1976 (p). Reprint of a special issue on women in the *Journal of American Folklore.*

Elizabeth W. Ferneau. *Guests of the Sheik: An Ethnography of an Iraqi Village.* Garden City, N.Y.: Doubleday, Anchor, 1965 (p). Revealing and sensitive study of the lives of Iraqi women under *purdah.* Surprising amount of diversity in their lives despite restrictiveness of this Islamic law.

Joan Goulianos, ed. *By a Woman Writt: Literature from Six Centuries by and about Women.* Baltimore: Penguin Books, 1973 (p). Good collection including some rare earlier pieces.

M. Esther Harding. *Woman's Mysteries: Ancient and Modern.* New York: G. P. Putnam's, 1971 (p). Jungian-inspired view of moon goddesses and so on.

Molly Haskell. *From Reverence to Rape: The Treatment of Women in the Movies.* Baltimore: Penguin Books, 1974 (p). How Hollywood portrays women – changes in films over a fifty-year period.

H. R. Hays. *The Dangerous Sex: The Myth of Feminine Evil.* 1964; rpt. New York: Pocket Books, 1966 (p). Enduring theme of the identification of women with evil, found in the myths, beliefs, and literature of almost all societies.

Carolyn G. Heilbrun. *Toward a Recognition of Androgyny.* New York: Harper & Row, 1973 (p). Another view of the sexes in literature this time emphasizing androgyny rather than sexual dualism.

Thomas Hess and Elizabeth C. Baker, eds. *Art and Sexual Politics: Why Have There Been No Great Women Artists?* New York: Macmillan Co., Collier Books, 1973 (p). Linda Nochlin's well-known essay on this question and ten replies by art historians.

Robin Lakoff. *Language and Woman's Place*. New York: Harper & Row, 1975 (p). How language reflects the sex role system.

Wolfgang Lederer. *The Fear of Women*. New York: Harcourt Brace Jovanovich, 1968 (p). Good presentation of symbolic material and an analysis of its relevance to sex roles and what the author terms "the fear of women." Some questionable conclusions.

Joan Mellen. *Women and Their Sexuality in the New Film*. New York: Dell Publishing Co., 1973 (p). How and why women are portrayed the way they are with an emphasis on recent trends.

Karen Petersen and J. J. Wilson. *Women Artists: Recognition and Reappraisal from the Early Middle Ages to the Twentieth Century*. New York: Harper & Row, 1976 (p). Good treatment of the lives and works of scores of forgotten women artists from both the West and the East.

Sarah B. Pomeroy. *Goddesses, Whores, Wives, and Slaves: Women in Classical Antiquity*. New York: Schocken Books, 1975 (p). Well-done study of women, real and mythic, in the classical world.

Eileen Power. *Medieval Women*. Cambridge: Cambridge University Press, 1975 (p). Interesting glimpse of women and the several roles they have played from the cloister to the village. Includes discussion on the cult of the Virgin and its significance.

Rosemary R. Ruether, ed. *Religion and Sexism: Images of Woman in the Jewish and Christian Traditions*. New York: Simon & Schuster, 1974 (p). Anthology of writings on the portrayal of women in the Bible and their role in religious life.

Patricia Meyer Spacks. *The Female Imagination*. New York: Knopf, 1975 (p). How women novelists have pictured the female condition.

George H. Tavard. *Woman in Christian Tradition*. Notre Dame: University of Indiana Press, 1973 (p). Theological background for the position of women in Christianity.

Barrie Thorne and Nancy Henley, eds. *Language and Sex: Difference and Dominance*. Rowley, Mass.: Newbury House Publishers, 1975 (p). Interesting articles and good bibliography.

Thorkil Vanggaard. *Phallos: A Symbol and Its History in the Male World*. New York: International Universities Press, 1972 (p). The use and significance of phallic imagery in art and in religion.

Marina Warner. *Alone of All Her Sex: The Myth and Cult of the Virgin Mary*. New York: Knopf, 1976. Historical study concerning the impact of the doctrine of the Virgin Mary on the position of women.

Virginia Woolf. *A Room of One's Own*. New York: Harcourt, Brace & World, 1957 (p). How social constraints keep women from creative work.

5

Sex Role Change through Space and Time

We have seen in the preceding chapters how well anchored the sex role system is in the biological, psychological, and social spheres. Yet recent years have seen an increasing number of programmatic efforts aimed at changing this system. How successful have these efforts been? And, more importantly, what does their success or failure tell us about the nature of the moorings of the sex role system? The tale has generally been one of some change with resistance to change occurring at several critical points. For example, the role and privileges of men have held fast while women's position has changed somewhat; moreover child-rearing and child care still have not moved outside the context of the nuclear family.

In this chapter, we will be concerned with change through space and through time. By the first, we mean a cross-cultural look at several societies that have instituted sex role changes as a matter of ideology and have introduced such changes with an institutional support system of a programmatic nature. In this category fall the societies of the Soviet Union and Eastern Europe, China, Israel, and Scandinavia (especially Sweden). In the second section of the chapter, we shall view the history of feminist thought, especially in the United States, in the perspective of what the future might hold for sex role change.

CHANGE THROUGH SPACE

All the societies to be discussed share one characteristic: that of socialism. To a greater or lesser degree, programs of sex role change have been based on the socialist dictum that only female participation in the labor force will lead to equality between the sexes. In the earlier chapter on the family, we reviewed the thought of Engels on the subject (see p. 128). To the Marxist way of thinking, the leisure-class housewife is the hallmark of bourgeois capitalist society—and a hallmark doomed to personal dissatisfaction at that, given the lack of useful purpose in her life. The practical necessities of building revolutionary societies, especially those that had not previously been industrialized (like the Soviet Union and China), require the full labor force participation of all citizens, adding an additional inducement for sex role change. The introduction of women into the labor force is not enough to produce change in the sex role *system,* however, as we shall see in our examination of the experiences of these varied societies.

THE U.S.S.R. AND EASTERN EUROPE

The Soviet Union, the first society to be based on Marxist principles, offers a particularly instructive case for the history of sex role change and resistance. The values and mores of prerevolutionary Russia were very supportive of ideas of male superiority. In his masterful book on Russia, the journalist Hedrick Smith quotes a few Russian proverbs: "A wife isn't a jug—she won't crack if you hit her a few," "When you take an eel by its tail or a woman by her word, there's precious little stays in your hands," and "A dog is wiser than a woman—he won't bark at his master."[1] Although Russia now has the highest percentage of working women (85 percent) in the industrialized world, the attitudes inherent in the proverbs still seep through to color female life. Although female emancipation was a very early stated goal of Russian Communism, a traveler in 1932 could report this village scene:

> the women stood while the men sat down and ate; kept their heads bent and their hands folded, not speaking until they were

> spoken to. In one hut when I asked the peasant woman a question, the husband repeated it to his wife, the wife answered him, and he returned the answer to me.[2]

In the present day, women are concentrated in the lowest-paid and lowest-status sectors of society. The images of the female construction worker and the female doctor do much to distort the Western view of Soviet sexual equality. To the superficial observer, the presence of women (in fact, the preponderance of women) in the construction trade bespeaks an open-minded liberalism of job fluidity independent of any antiquated ideas of fields that should be "naturally" closed to women because of their physical weakness. When we take a closer look, we see something very different. Heavy manual labor is on the lowest rung of the status and income ladder and is viewed as "women's work" in a derisive way. Far from being liberated to follow such occupations, women are often limited to them. World War II, with its tremendous loss of Soviet men, provided the population pressure necessary to accelerate the siphoning off of women into previously male jobs. After the male population dip leveled off in the postwar years, the patterns remained since they were consonant with the prevailing low estimation of women and their image as "beasts of burden." So, although the Marxist precept of female labor force participation has been followed to the letter, its spirit has obviously been forgotten: instead it is that of prerevolutionary Russia.

But, the skeptical reader is surely saying, the case of female Soviet physicians cannot be dismissed so easily. Everyone knows that three-quarters of Russia's doctors are women; is that not the case? It is, but what is implied by that fact is quite different from what equivalent statistics would mean in the United States. Medicine is a low-paying job with little prestige, and therefore is consonant with the actual (if not the ideal) image of women. The Soviet statistics on medicine include many workers with less skill and lower status than physicians in our society. It is in these lower realms that women predominate. Surgery and the higher administrative posts in medical schools and hospitals remain primarily male domains. Male university graduates choose the sciences or industry, where pay and prestige are better, leaving medicine open to women. The average physician in the nationalized health plans of Eastern Europe makes less than the industrial production worker. Again, as in the case of heavy construction, the decline in the male population brought about by World War II provided the initial opening for women and accelerated their entry into medicine. By the 1960s, however, the tide was turning, and fears about the "feminization" and consequent degrading of medicine appeared. As a result, it became more difficult for women to enter medical school. A Czech economist wrote of this policy in 1966:

> The apparition of feminization has emerged suddenly in horrible guise, stalking the colleges and secondary schools, the relevant organs of governments and institutions. And horrifying everyone. Thy name is woman, and therefore know thee that thou must relinquish any desire for medicine, teaching, philology, sociology, and I don't know what else. Some years ago we were carried away by the number of callings which had been invaded by women and in which they had proved worthy. Now we are carefully, pedantically weighing and choosing one occupation after another which women must not be allowed to enter.[3]

Similarly, in industry and on collective farms, women are rarely found at the management level, even in cases where the majority of workers are women. The situation with regard to high governmental positions is comparable. As Smith points out, Russian journalists often brag that more women are found in the Supreme Soviet "than in all the parliaments of the capitalist states combined," but they do not add that this group is just a "rubber stamp" body with no real power of its own. The Communist Party apparatus in which the real power resides is almost totally a male domain at the upper levels. Smith also points out that even at provincial levels, few women hold positions that would be comparable to our governorships. As a final indication of what Smith terms "unselfconscious male chauvinism," he points to the fact that the Soviet Commission for International Women's Year (1975) was headed by a man.[4]

We can see then that the residual effects of traditional attitudes far outweigh the pronouncements of official egalitarian ideology, especially when promulgated and implemented by individuals (of both sexes) who are steeped in the folk beliefs of the culture. On one level the widespread entry of women into the work force seems to satisfy the Marxist tenet of female liberation, but in reality, it only lifts the struggle to a higher level. Women are still the second sex in Russia, but they now play that role in the workplace as well as in the home and on the farm.

The institutional arrangements that permit female employment point out some of the inherent difficulties encountered when the female sex role is altered without corresponding (and complementary) changes in the male role. Day care facilities do exist—usually at the industrial site—but they are not totally adequate. Older female relatives, often the grandmother when available, take up the slack and provide more personalized child care. In all cases, though, the responsibility for child care as well as for household maintenance falls exclusively to the female. Sharing of housework and child care is practically unknown, and the institutional supports (such as paternity leave) do not exist, nor would they be acceptable in the climate of thought common in Russia and Eastern Europe. Women's work in the home and for children is just that, women's work, given a lower order of status and encompassing a domain not to be touched by any self-respecting male. The result, of course, is that women find they have two roles, housewife and worker, to the male's one, worker, with only one, worker, receiving any societal or monetary recognition. The problem of "equality" within these bounds is expressed by a member of the Czech women's national parachute jumping team:

> Our parachutes are the same as theirs, we jump from the same planes, we've got guts, and our performance isn't much different from the men's, but that's where emancipation ends. I'm married, I'm employed, and I have a daughter. And a granny. If I couldn't say: "Granny, keep an eye on her," my sporting career would be at an end.[5]

When children are ill, it is the mother who stays at home, and it is the societal expectation that she do so. A Czech survey showed that over a third of working women would stay home permanently if they could, and the percentage increased among women who had less prestigious, lower-paying jobs. Minimal standards of living often cannot be maintained, however, unless both husband and wife work, so that economic pressure as well as societal climate funnel many women into the labor force.

Women are responding to the pressures of job and home by having fewer children, resulting in what Hilda Scott terms an "underpopulation crisis." Abortion seems to be the most common method of birth control, as other means are not readily available or widely used. The sexual prudishness characteristic of the Soviet Union and much of Eastern Europe also doubtless contributes to this situation. Instances of women having a half-dozen or more abortions are not uncommon. The desire to have children is there, it is just that the practical reality of a working mother's having to raise them with a minimum of outside support militates against multiple-child families. Smith reports a 1973 survey of 33,000 Soviet women that found that "the vast majority of women want two or three times as many children as they actually have." Smith further reports on a survey that found that while 3 percent of women defined having only one child as ideal, 64 percent did in fact have only one child and 17 percent had no children.[6] This child deficit flies in the face of numerous Soviet campaigns (including awarding medals to mothers of many children) to encourage childbearing to replace the population lost during World War II. The economic difficulties of raising several children on minimal wages also contribute to this situation, but it is clear that the strain on the Soviet woman from her two roles is a major factor.

The situation in the Soviet Union, then, and in other Communist countries of Eastern Europe is one of modified sex role change. The value system of male supremacy has been left virtually untouched although women have entered the labor force in increasing numbers. The occupations females find themselves in tend to be at the bottom of the income and status hierarchies. Women do not play any real (as apart from ceremonial) role in the political system; they are also excluded from active entry into the military. More important, the traditionally female domain of housework and child care has remained the sphere of women; men have not entered these realms either physically or psychologically. No real change has occurred in the roles of men; they have merely looked on while women have taken on employment. The essential attitudes and balance of power between the sexes remains unchanged. The sex role system as it existed before the Russian Revolution remains virtually intact in all its important structural features although some of the window dressing looks a bit different.

The Soviet and Eastern European cases demonstrate that the mere (even mass) entry of women into the labor force will not demonstrably alter the sex role system and the balance of power. Rather (as in the United States), the employment profile of women and the status of jobs will change shape to fit the powerful attitudes and structures of the traditional sex role system. When women hold no real political or social power and are regarded with the same attitudes as before, and (most important) when male sex roles change in no real way, the sex role system stands solid. When male domains of power and prestige and female domains of home and child care exist as physical and psychological realities, the lines of sex role structure remain inviolate regardless of ideological claims to the contrary.

CHINA

But, the skeptical reader is surely thinking, what of the revolutionary society of China? Much of the recent flurry of writing and publicity about China has centered on the new role of women.

Much of the writing has also dwelt on the contrast between the extreme prerevolutionary oppression of women and their modern opportunities and options. Seldom has a society been as patriarchal as prerevolutionary China. Women were strictly segregated from public life, were left uneducated and seen fit only to endure a difficult life of household duties and child care. Female children were viewed as less desirable than male: the ideal was to have sons to carry on the family name. Mencius, the disciple of Confucius, said, "There are three unfilial acts; the greatest of these is the failure to produce sons."[7] A woman's life was dominated by her relationship to men and her position in the extended family. The undisputed family head was male and any power the woman had derived from her association with him, although this association was most often one of unquestioning obedience rather than transferred authority. Girls were commonly married in their late teens through arranged marriages. They then came under the authority of their husband and mother-in-law. The tyranny of the mother-in-law over the new daughter-in-law is a well-known image in Chinese culture and can be partially explained as being the only real aggressive outlet sanctioned for women. It is likely that all the pent-up emotions of dissatisfaction and hatred against the patriarchal family system were channelled into this one permitted avenue for expression of hostilities and exercise of authority. Ironically, though, the target was, like the mother-in-law herself, only another helpless victim of the system. Doubtless, the harsh treatment of the daughter-in-law provided a modeling experience for the young woman who then bided her time until she was permitted this outlet for her frustrations on her own daughter-in-law.

The low valuation of women was often expressed most directly in the common practice of female infanticide. While sons were eagerly awaited as avatars of the family name, daughters were only tolerated if enough economic resources were available to maintain their support while they waited for a strategic betrothal to advance the patriarchal lineage. Fu Hsuan, the Chinese poet, comments on women's fate in a (truly) man's world:

> How sad it is to be a woman!
> Nothing on earth is held so cheap.
> Boys stand leaning at the door
> Like Gods fallen out of heaven.
> Their hearts brave the Four Oceans,
> The wind and dust of a thousand miles.
> No one is glad when a girl is born;
> By her the family sets no store.[8]

Women were often abused physically as well as psychologically within the family and wife-beating was an accepted (and expected) practice. No avenue of escape was provided, for female-initiated divorce was unheard of in traditional China. In any case a woman could have no real social or economic existence outside of the family, her only recognized sphere. Left uneducated and unprepared for any other future, she was condemned to her familial place.

In addition to housework and child care, woman's task was also sexual, but sexual in the direction of satisfying male needs. Taoism (as we saw in the last chapter) emphasized the mutual need of each sex for the sexual essence of the other, transmitted through sexual intercourse. In practice, however, the sexual life of the male was taken as much more central. Men could freely take concubines into their household (if they could afford them), adding a fresh sexual source for men. Women, of course, were permitted no such outlet if their marital life proved unsatisfactory. The notorious Chinese practice of foot-binding, in which young girls had their feet bound tightly at any early age to prevent growth, was developed because of the erotic appeal of the so-called "Golden Lotus" foot. This appeal was, of course, to men, as the afflicted women could not walk on three-inch-long feet and could not be expected to recall with joy the painful childhood days of the foot-binding. This practice, said to have originated in 937 A.D., persisted until the 1940s (although it was officially outlawed in 1911) and was mainly limited to the upper classes, since, of course, the lower classes could not afford the lost labor involved in the invalided and immobile women. Since foot-bound women had to be carried around, their physical imprisonment mirrored their psychological one, both within the strict confines of the Chinese patriarchal family system.

It is necessary to understand the "bitter past" of Chinese women to appreciate the changes that have been made and to grasp their limited nature. Not all the changes that have been introduced came after the Communist takeover. The opening of China to the West and the Republican Revolution of 1911 brought many changes, but these were mainly limited to the upper classes. For example, "natural foot" societies arose and pressure was brought both legally and socially against the practice of foot-binding. As in the present day, women in the cities were first to take advantage of the new freedoms; change in the rural countryside (where the large majority of Chinese live) lagged far behind. Before the Communist takeover, women had already begun a limited entry into public life, gaining education and some occupational experience (in female jobs like teaching and nursing). The practices of arranged marriage and child betrothal were also lessening, although, again, such changes reached only a very small proportion of the population.

Women were not very active in the actual Chinese Revolution. Curtin reports that only fifty women participated in the famous Long March of 1934 and 1935, out of a total of well over 100,000 Chinese on the march. Even those few women were only to be found in special areas of the army, notably the health corps. After the Communist victory in 1949, however, the liberation of women was given a much more central role. Perhaps it is incorrect to view the movement as one primarily aimed at life change for women, since the overall goal of overthrowing the authority of the patriarchal family and the feudal land system seems more critical to the Communist goal of replacing reverence for past tradition with adherence to present and future Marxist doctrine. The most direct way to accomplish this end was to break up the family. As an additional dividend of this breakup, the allegiance of one half of China's population, the women, was assured by offering them a way out of the oppressive family system. The authority of the state might have seemed a beneficient alternative to that of the patriarch and mother-in-law. Education and labor participation were also stressed and literacy and economic independence accelerated the pace of female entry into society. Such entry was, of course, also required for the rapid industrialization and economic development of China.

The Communists created "speak bitterness" sessions in which women were encouraged to recount the horrors of their prerevolutionary existence through the means of group support and interaction. Present-day American "consciousness raising" groups operate on the same principles and often achieve the same results: recognition that individual unhappiness is a result of societal and familial structure and is a shared female experience that can provide the basis for future cooperative change. Women were not encouraged to turn against men, but rather against the prerevolutionary patriarchal feudal system, which was the ultimate target of the Communists. Men, particularly the majority poor, were also oppressed by the landlord system and met in similar sessions to vent their feelings and be presented with the Communist program for change. Considerable physical and psychological coercion was present in the system of thought reform, and those who did find satisfaction with the old system could find no audience for their views, and indeed would have been severely sanctioned for expressing them if they dared.

Indeed, it is useful to mention here the difficulty of ascertaining the degree of dissent or diversity in present-day China. Until very recently few Westerners had access to China, and even now that access is tightly regulated. Visitors are closely supervised, and because of this and the language barrier, the view they get of Chinese society corresponds very closely to that of the ideal projected by the Communist government. In addition, of course, severe sanctions are imposed against Chinese critics, and this can account for the seeming sunny uniformity of opinion. Our knowledge of the Soviet Union is similarly circumscribed, but not to the degree it is in the Chinese case. Therefore, accounts brought out by Westerners (even those not overtly sympathetic to the Communist regime) must be read with these caveats in mind.

The knowledge we do have of present-day China indicates that women's position has been reformed—not really revolutionized. The authority of the patriarchal family seems to have been greatly reduced, but the persistence of many traditional sex role ideas is notable. We hear little today of the tyranny of the mother-in-law—daughter-in-law relationship and of cruelty in the marital bond. The Marriage Law of 1950 mandated free choice of marriage partners and made divorce easily available. Child betrothal was outlawed and the principle of equality in marriage was established. Much resistance against the law was noted, especially in the more conservative countryside, and many of the party cadres sent out to enforce it did so only half-heartedly.[9] However, many changes were put into effect, and the number of divorces—almost all brought by women—increased drastically. Women who sought divorces had the psychological support of the party organization and the chance to exist outside of the family, an option not readily available before. In the ensuing twenty-five years, the government has taken a harder line on divorce, and now cases of marital discord are arbitrated before a communal meeting and the couples are expected to remain together. Companionship and joint dedication to party goals are stressed as cement for marital ties. Since marriage choice is now free, at least some of the reason for divorce would presumably be gone. Individual happiness and fulfillment and the accumulation of material goods are not cultural goals in the sense that they are in our society. Sexuality is seen as a natural function, but its exercise outside of marriage is frowned upon. Women and men dress alike in unadorned clothing, giving little room for sexual display, and, of course, since China has no consumer advertising, sex in the public domain is essentially absent. Women are not viewed as sex objects, at least not publicly, and the entire sexual tone of the society is decidedly muted. What role the sexual vision of femininity plays in our own sex role system is difficult to determine, but it surely does contribute to the secondary status of the second sex and the one-dimensional view of womanhood that maintains that status.

However, even with the considerable changes that have been effected in China, some enduring continuities persist. While women now get a few years of education and almost all can read and write, relatively few go on to the higher levels of education attained by men. An American visitor to China questioned the disparity between the sexes in the junior middle school (only 20 percent women) as compared with their parity (50 percent) in the primary school. She was first told that many more boys than girls had been born in the region. Skeptical about that reasoning, she finally was given another explanation: older girls were expected to help with the housework.[10] And, indeed, housework and child care have remained exclusively female domains. Communal nurseries are entirely staffed by women, as are teaching posts in the lower elementary grades. Mothers are expected to work and day care facilities for children are usually provided at the workplace (almost always at the *mother's* workplace). A sizable proportion of children with working mothers (and about 90 percent of mothers do work) are cared for by female relatives, usually grandmothers, or by female neighbors. The official Chinese policy is not aimed at altering these practices, and Western visitors often encounter the attitude that such sex-delimited activities are "natural," and that one should expect that women and men would enter different occupations and be differentially involved in child care.

While many Chinese women are involved at local levels of government, few are found at the higher levels of the regional or national administration. Likewise, supervisory posts in industrial plants and communes are almost always exclusively held by males. Women have entered medicine, but mostly the fields of pediatrics, obstetrics/gynecology, and psychiatry. Many of the "barefoot doctors," who have been given minimal medical education and minister to peasants in the outlying provinces, are female. Some wage disparities exist, especially on communes, where women are paid less because of their household work (which presumably would detract from their communal contribution) and their "weakness" and inability to do many of the harder physical tasks.

Finally, we might quote Kate Curtin's retelling of the experience of a group of Westerners in China:

The sexual determination of roles is most pronounced in the

unpaid household labor. There the patriarchal division of labor inherited from the old system has been most resistant to change. The Committee of Concerned Asian Scholars report that everywhere in China they asked questions about who cleaned house, washed clothes, took care of children, did the shopping, cooked the meals, did the sewing. The answer was almost invariably "The wife, of course."[11]

In short, then, we find little structural change in the system of sex roles, although undoubtedly the contrast with the prior misery of Chinese women in the patriarchal family makes the progress seem more impressive than it is. The allocation of male and female work and family domains remains inviolate. The cultural belief in the different psychological orientations of the sexes is strong, but is not as obvious in a society where no leisure class of housewives exists and in which the Madison Avenue image of the feminine mystique could serve no function. The power structure of the society is male, as is that of the military, a most important component in any totalitarian regime. The lives of individual women have changed dramatically: literacy, freedom from immediate familial domination and tyranny, and options for work outside of the home have been real advances. Yet in many ways Chinese women are at the same point as women in the United States, having come to it from a totally different direction. In China, as in the United States, men do not participate in housework or child care, at least not to any noticeable degree, and are not expected to by the society. As in the Soviet Union, women work in the home and outside of it. In China, the birthrate has also dropped, although this may be in part a result of the concerted government campaign to cut down on overpopulation. We hear little individual or group dissatisfaction expressed by Chinese women, but of course the government does not permit dissent or considerations of individual happiness over group goals. The search for individual fulfillment, an idea readily understood by any American is not acceptable in China. In any case, the contrast between the "bitter past" of most Chinese women and men and the more livable current scene is likely to militate against too stringent a critical examination of the present social structure, let alone the sex role system embedded within it.

We have seen, then, that in both the Soviet Union and China, the lives of women have been substantially changed without any real effects on the structure of the sex role system. Men's roles have not changed, and women remain psychologically in the home even while spending a large part of the day out of it. Is our search for a revolutionary sex role society doomed? We press on, to two Western societies that have instituted programmatic efforts at sex role change: Israel and Sweden.

ISRAEL

The Israeli case is one in which only a small minority of the population has been involved: the estimated 4 percent who live on kibbutzim. However, we shall consider the kibbutz in the broader perspective of Israeli value patterns since they have had a decisive influence on the outcome of the kibbutz experiment. The kibbutzim were begun and mainly peopled by Jews from Russia and Eastern Europe who brought socialist ideas with them to Palestine (for the kibbutzim antedated the birth of Israel in 1948). It is therefore instructive to note the relative roles of women and men in the shtetl culture from which the vast majority of the kibbutz immigrants came. Religious and public life (insofar as it was permitted to Jews) was the domain of men; women were in charge of the home, although they often contributed to the support of the family. The ideal case, however, was one in which the husband spent as much time as possible in Torah and Talmudic study and the wife kept a home full of emotional warmth and good food. The mother-child link was taken extremely seriously, and a woman's ideal goal was to further the happiness of her children and husband. Zborowski and Herzog, in their fine book on the shtetl, *Life Is With People,* comment:

> The woman's informal status is more demanding and more rewarding than that formally assigned to her, for in actual living the complementary character of her role is always to the fore. She is the wife, who orders the functioning of the household and provides the setting in which each member performs his part. She is the mother, key figure in the family constellation. Moreover, the more completely her husband fulfills the ideal picture of the man as scholar, the more essential is the wife as realist and mediator between his ivory tower and the hurly-burly of everyday life.[12]

The image of Eastern European Jewish wife and mother implies considerably more importance and centrality than our American case of the feminine-mystique-influenced housewife of leisure and little purpose. However, they share an invisibility within the home, a hidden status outside the societal ranking system. Socially recognized power and regard went to men, particularly those men who could achieve the coveted goal of rabbinical scholarship. Within the culture there was no tradition of female participation or leadership outside the home. Although many of the kibbutz immigrants came from the cities, where the patterns were slightly less role-segregated, none was very far removed from her or his shtetl past.

The other major group of Jewish immigrants to Israel, the Sephardic Jews from the Middle East (Yemen, Syria, etc.), brought a similar tradition of sex role segregation to their new Israeli home. Although relatively few Sephardic Jews populated the kibbutzim, their values helped shape the general Israeli sex role climate. If anything, Jews from the Middle East brought even more rigid ideas of sex role division with them. Influenced by their Islamic and Arab neighbors, they accepted many of the prevailing notions of sexual segregation and the low status of women, emphasizing the withdrawal of women from public life. The family structure could be characterized as strictly patriarchal, with authority emanating from the male elders.

It was with these backgrounds that the kibbutz pioneers aimed to bring about a sex role revolution. The overall emphasis was on a "return to the soil," for it was thought that only by abandoning traditional Jewish occupations of commerce and trade could a new nation be built and a new generation of truly free Jews created. The agricultural settlements were built on socialist principles (the first were begun at the same time collective farms got their start in the Soviet Union), including equal sharing of wealth and the elimination of private property. The kibbutz itself rather than the family was to provide for every need and to take over many of the traditional functions of the family. The latter aspect was undertaken with the twin aims of encouraging greater loyalty to the community unit rather than to the private consuming family and of making possible the emancipation of women from domestic tasks, a goal derived from Engels' analysis of the family (see p. 128).

On the kibbutzim, household work and child care were made a common responsibility, with communal kitchens, minimal home settings, and collective care of children, thereby "freeing" women from the drudgery of running their own homes and raising their own children, presumably the root cause of their dissatisfaction and a prime symptom of the decadence of capitalist societies. However, it seems that almost from the start the so-called "public domain" that household and child care tasks were put into was in reality a *female* domain. Women became the *metaplot* (caretakers) and the service workers in the kitchens, while men worked in the fields and took over the supervisory positions. Spiro states the case nicely: "instead of cooking and sewing and baking and cleaning and laundering and caring for children, the kibbutz woman cooks *or* launders *or* takes care of children for eight hours a day. She has become a specialist in one aspect of housekeeping. But this new housekeeping is more boring and less rewarding than the traditional type."[13]

The critical issue here seems to be the same as in the Soviet Union and China, the retention of traditional sex role ideas within a framework of greater labor force participation by women. Given the special case of the kibbutz, with its collectivization of domestic functions, the sex role division of labor becomes almost a caricature of the traditional roles. A study done in 1967 found that most men on the kibbutzim worked in traditionally masculine occupations (between 53 and 84 percent), while most women were engaged in feminine occupations (between 50 and 84 percent).[14] The absence of men in female occupations (such as that of *metapelet*) was particularly notable: in ten of the eighteen kibbutzim studied, no men were in feminine jobs. What seems to have occurred on the kibbutz, according to Rabin, is a "masculinization" of kibbutz roles. Personal adornment, sex-distinctive clothing, and social etiquette were devalued, and the masculine values of physical strength and aggressive leadership came to be taken as ideals for the kibbutz as a whole. In such a climate, female occupations, which did not result in economic reward for the group, were devalued, and, implicitly, so were the women who occupied such roles.

The question arises as to why the sexual division of labor was allowed to become the norm, given the dogmatic assertions of sexual equality that preceded the founding of the kibbutz movement. One possible answer is that separate sexual realms of labor were thought to be "natural" and not incompatible with sexual equality since both agricultural and service occupations were essential to the running of the kibbutz. Of course in this case, separate was unequal, given the generally low regard for household labor and the difficulty of assessing its value in readily understood monetary terms. Another answer is that men were simply unwilling to take on traditionally female roles, thus leaving them by default to the women. Given the background of the men and their previous noninvolvement with household and child-rearing tasks, this attitude is comprehensible. The women themselves may have found it difficult to abandon their old values or simply may not have recognized the implications of the separation of realms. Also, of course, the primitive agricultural setting in which clearing land was necessary was one that maximized the importance of differences in physical size and strength. Men may have been simply more efficient in doing such tasks. Given the precarious geographic setting of Israel, agricultural work was almost indistinguishable from defense work, since hostile raids on kibbutz borders were not at all unusual. From the start, men were the military defenders of Israel and still are (women occupy only noncombat roles and serve a much more limited term of service). The aggressive military spirit required for settlement amid hostile neighbors contributed to the widening split between the sexes. In such a setting, attributes traditionally associated with masculinity, such as aggression, became much more important than the more traditionally feminine values of domesticity and emotional warmth that might be more prominent in more secure societies.

In any case, Rabin identifies as one of the key problems of the kibbutz what he terms "the problem of the woman." He asserts that "whenever a family leaves the Kibbutz it is usually the woman who is the main instigator of the move."[15] The dissatisfaction of women with their job opportunities on the kibbutz appears to be only one reason, albeit an important one. Men for the most part started the kibbutz movement and continue on in management roles; hence there is resistance to the necessary changes that would give more options to women (mainly male unwillingness to do "female" work). The kibbutz women, it might be added, also adhere to the traditional notions of sex-typed occupations.

Another, and perhaps more revealing, source of female dissatisfaction is with the communal care of children. Surveys and informal reports seem to indicate that mothers simply want to spend more time with their children, even to the extent of wishing for a return to the traditional nuclear family with full-time care by the mother! The *New York Times*[16] reported a survey that found that two-thirds of the women and one-half of the men opposed the traditional kibbutz system of children living apart from their parents except for a few hours' visiting per day. Some kibbutzim have already returned to modified or traditional forms of the nuclear family for the raising of children. The same article quoted a kibbutz woman: "Kibbutz women aren't interested in equality, they're interested in children." Many kibbutzim have instituted a so-called "hour of love," a time before work when mothers may be with their children (in addition to a three-hour period after work). The *Times* reports: "The younger women instigated the 'hour of love' against strong opposition in the 60s, and defend it with feline intensity. It takes top priority: even if they get the chance, young mothers refuse work in the fields or outside the kibbutz so that they can be available for it. Like the stereotype of the Jewish mother, they center their lives around their children." Such vignettes and more thorough studies that substantiate such images have prompted biologically-oriented anthropologists like Lionel Tiger to posit an innate mother-child bond that resists the onslaught of ideological superstructures. However, other explanations also come to mind. Given the lack of career opportunity for women on the kibbutz and the Eastern European tradition of mother-centeredness, motherhood may simply provide the most rewarding role for women on the kibbutz. The masculine military values that infuse kibbutz life offer little other outlet for more "feminine" values, values that may have some biological base but that surely have been inculcated by the traditional Jewish background of the kibbutz pioneers. Kibbutz women also seem to be responsive to other aspects of what we would call the "feminine mystique." The *Times* quotes two "popular myths of kibbutz women": "Working outdoors ages your skin prematurely" and "driving a tractor makes you infertile."

The necessities of child care seem to have pushed the balance toward the traditional division of sex roles. Female jobs on the kibbutz were low in status, and male values of aggression reigned supreme; hence women were left with no socially recognized role. The *Times* comments on the younger women's response to this situation:

> The younger women reacted to their mothers' inevitable failure and loss of status by falling back on an assured method of production: Reproduction. Bearing and rearing children is the most valuable female role on the kibbutz and one in which women could regain productive status.[17]

One might also note that the Israeli climate of thought about sex roles actively fosters this sexual division of labor. Stiller reports that Israeli commercials are decidedly sexist, with domestic products for women being the featured items. Women in such commercials are portrayed as happily enjoying their homemaking role. Stiller also notes:

> Girl-watching of the most obvious kind is a favorite Israeli sport. . . . A powerful streak of what can only be called male chauvinism is present in Israeli social relations. Its manifestations are not subtle and are readily detected by any fairly attractive woman who walks down a street, stops for a cup of coffee,

looks at a newstand (where girlie magazines abound), or goes to the movies.[18]

She goes on to report that women are encouraged to think in terms of early marriage and childbearing and to value domesticity. Stiller comments: "While I am not advocating domestic negligence, it is pretty obvious that the incessant and overwhelming attention paid to how a house is kept and how a cake is baked stems from an image of self."[19] Occupational stratification by sex is quite rigid, and domestic chores and child-rearing are almost never undertaken by men, both as a matter of pride as well as convenience. (Stiller wryly notes: "There may be a Moroccan somewhere in Israel who washes the floor for his wife once in a while, but I have yet to hear him admit it publicly."[20] The sex-role-traditional backgrounds of both the Sephardic and Ashkenasic portions of the population have had their influence, but they have been taken out of their original cultural context. For example, Stiller points out:

> What is most notable in this (Eastern European) group is the transference of certain middle and eastern European values from mothers to daughters, without the backbone of belief which may have spiritualized the lives of women in the towns and *shtetls* of the old country. The Friday *challah* [ceremonial bread], for example, seems to be important not so much for its religious symbolism, but as proof of how good a *balabustah* [housekeeper] the woman is.[21]

In her analysis, Stiller points to two other root causes of what she calls the "battle of the sexes" in Israel. One is the orthodox religious establishment that enforces the marriage laws that give women a secondary place. For example, women may not initiate divorce suits in Israel, as this is forbidden under orthodox Jewish rite, the only one recognized for Jews in Israel. Israeli women must also be certified as widows in order to be permitted remarriage. This certification requires a witness to the husband's death, which is often difficult in combat situations. Jews must also be married under orthodox rite in Israel, and as we have seen in the preceding chapter, Jewish marriage laws, such as those concerning *niddah* or separation during menstruation, embody a negative image of woman. In any case, many of these rites and laws are only formalities that are not strictly adhered to, but they do act to enhance the traditional atmosphere of sex role segregation and female inferiority.

Perhaps a more compelling contributor to Israeli attitudes about sex roles is the almost continuous state of war and military readiness that has been Israel's fate since its creation. Stiller perceptively comments:

> At the most crucial moments of their lives, in those moments when they are fighting for their survival, the men remain with other men. The women in the army provide secretarial skills and moral support, but it is the men, finally, who go out on the field to die. . . . Men and women feel that the important work of defending the country is both a man's bane and his superiority. There is a regression to an older social code: men as defenders, the women as emotional supporters.[22]

Aside from the special case of Israel's need for military preparedness, which helps to tip the balance toward sex role traditionalism, many of the same themes have emerged here as in the previous cases of the Soviet Union and China. We see the relative rigidity of male roles and the common perception of differing psychological orientations leading to what we can only call a pattern of "woman's two roles," after a well-known book by Myrdal and Klein published in 1956.[23] In short, women are accepted (or even pushed) into the labor force, but their household and child care duties remain with only slight modifications, given men's unwillingness to share domestic duties (or the structural difficulty of allotting such work to men). And of women's two roles, the traditional domestic sphere always has priority: a woman is mother and wife first, worker second, both psychologically and sociologically. On the kibbutz, the nature of the second role is often similar to that of the first. "Work" usually means child care or domestic duties, or in any case, nurturant ones, such as nursing, and in almost all cases such jobs are accorded lower status than male occupations. In addition, the female position in the work world is reflected in her nonstatus in the public and political sphere: with few exceptions, men predominate in power and decision-making positions.

An important lesson from the above instances is that women's roles cannot *really* change if those of men do not, which implies that domestic as well as work roles must be shared on a sex-free basis. When that is the case, it will pave the way toward changing female life plans to eliminate "until" work, jobs taken "until" marriage or child care, or jobs taken that can be compatible with breaks for child care and household work (which necessarily limit occupational advancement). In addition, of course, much of the traditional sex role thinking about different psychological spheres for men and women derives from the identification of the female with the home and children, and if this tie could be broken, much of the strength of the attitudinal association would vanish as well. Of course, such biologically oriented thinkers as Lionel Tiger[24] have argued that the mother-child link is an innate one that will be extremely resistant to breakage, but in the absence of societal alternatives to the contrary and less traditional socialization models we cannot assign causality so casually.

SWEDEN

It is in the case of Scandinavia and particularly Sweden that the analysis of change in women's roles has reached the point outlined above. There has been a transition from the more usual model of changing women's roles to accommodate work (along with the traditional domestic and child care chores) to a model of *sex role change*, where both male and female roles are radically modified, especially in the critical area of reassigning housework and child care to both sexes rather than to the working mother alone. Before exploring this problem and the outcome of these attempts at radical reform, we shall fill in some background on the history of sex roles in Scandinavia.

In the nineteenth century, the Scandinavian countries were among the leaders in feminist and social reform movements. The modern feminist thinking of Ibsen in *A Doll's House*, where a woman is portrayed as rebelling against the constrictions of bourgeois marriage, is indicative of much of the thinking of the time. In 1915 a book appeared entitled *Feminism in Germany and Scandinavia,* in which the author, Katharine Anthony, favorably contrasted the Scandinavian women with their less independent German counterparts. She even notes the mythological Valkyries, strong images of Scandinavian womanhood:

> Much more than the German woman, has the Scandinavian woman preserved the traditional freedom and power of their common Teutonic ancestress. In the history of the Scandinavian wars lies one of the reasons. During long periods, sometimes for years, the men were withdrawn from the northern peninsulas and the women were left alone to keep up their homes and maintain the country. The independence thus cultivated in the women did not, of course, take flight upon the return of the ruling sex. The practice of authority and responsibility can not be so quickly unlearned.[25]

Whether Anthony's reasoning is accurate is difficult to determine, but it is clear that Scandinavia was in the vanguard of feminist reform in the late nineteenth and early twentieth centuries. Norway, in 1901, was the first European nation to give (limited) suffrage to women and was quickly followed by the other Scandinavian nations. But, of course, as we can see from our own history and our current situation, granting of the vote and reforming sex roles are very different issues.

In Sweden until the early 1960s, a moderate view of sex role change predominated in which no radical alteration in existing roles for men or domestic roles for women was envisioned; in short, the "two roles" view we have been examining predominated as an unsatisfactory "solution." As has happened in all industrializing countries (and Scandinavia is still heavily agrarian), more women were joining the labor force, but not being relieved of their domestic duties; and the problems of child care and the onerousness of the two roles remained. Occupational stratification by sex was strong and attitudes linking women to domesticity and childbearing and secondary job status were the norm. Male role change was not considered relevant, or perhaps more accurately, was not considered at all. Then, in 1962, Eva Moberg published an article, "The Conditional Emancipation of Women," that foresaw the necessity for change in the roles of both sexes and a radical reallotment of domestic responsibilities as prerequisites for any real change in the position of women. She wrote:

> The concept double roles can have an unhappy effect in the long run. It perpetuates the idea that woman has an inherent main task, the care and upbringing of children, homemaking and keeping the family together. . . . Both men and women have one main role, that of being human beings. . . . [What we have now is] conditional emancipation. Woman has been made free only under the unspoken condition that she still sees as her main task the care and upbringing of children and the creation of their environment. Just so long as she understands that this is her natural task, somehow built into her as a member of the female sex, then, of course society recognizes her not as a fully free individual.[26]

Beginning in the early 1960s, Swedish social scientists and feminist reformers changed their analysis and plans for change to encompass the entire sex role system, particularly that part concerned with child care. Soon after, psychological and sociological studies appeared that showed the extent of sex role socialization in the home, schools, textbooks, and the media. Male as well as female stereotypes were analyzed, and the need to revise attitudes toward occupational choice was recognized. Most important, legislation was enacted that recommended that child care be a responsibility of both parents. This legislation provided for extensive child care centers and parenthood benefits, including a guaranteed income for six months after a child's birth for either parent caring for the child and sick pay allowance for a parent staying at home with a sick child. Ironically, Sweden has one of the lowest birth rates in the world (15.8 in 1966) and one-child families are common. Yet, as we have seen in our analysis of the American case, the realities of child care and the attitudinal effect of anticipated child care can be quite different, with the latter often having a far more pervasive influence than the former. In Sweden the age at marriage has also dropped (the average age for women was 22.9 in 1960, compared with 27.8 in 1930), so that child-care-free years have now increased. In any case, the problem of caring for children does loom large attitudinally as well as practically in the early years of female careers, and much legislation has been aimed at officially freeing women from compulsory child care. Still, in 1975 it was estimated that only 34 percent of the needed day care facilities were available, and extended hours for schools were being proposed as an alternative. About three-quarters of Swedish women with school-age children do work, although as we shall see, patterns of occupational choice are a key to the tenacity of the old attitudes in the face of new legislation. In 1966, 57 percent of women who did not work outside the home gave as their reason either household duties or child care.

The official attitude of Sweden toward sex role reform has received wide attention and is best known from a 1968 report to the United Nations, "The Status of Women in Sweden." In that report, the conventional view of economic and domestic functions within the family was attacked as a key to the sex role revolution:

> The idea that women must be financially supported by marriage must effectively be opposed—even in law, as it is a direct obstacle to woman's economic independence and her chances to compete on equal footing with men in the labor market. By the same token, the traditional duty of the man to support his wife must be supplanted by responsibility, shared with her, for the support of the children. This support of the children should also find expression, from the man's side, in a greater share in the care and upbringing of the children.[27]

Numerous government panels have also been commissioned to examine sex stereotyping practices in schooling and texts, particularly in the critical area of occupational choice. Yet the tale is not one of success, but rather of the retention of old attitudes, even in a country with a very low birth rate and liberal views on marriage and sex. In the case of child care there is a substantial unofficial view that opposes communal care for children. A critic of the official position, Inger Rudberg, voices her view:

> The question of the care and upbringing of children is the crux of the whole sex role debate, which should really be called the child debate, since it is the children who in the last resort are most concerned. Not all parents regard day nurseries as the ideal way of caring for their children. They think that for the mother, with the father's economic support and interested assistance, chiefly to take care of her children represents an irreplaceable value. This is a view and an evaluation which should be respected. It is not so much a question of the economically measurable uses of a mother in the home; it is a question of quite other values—of her personal contact and being together with her children, giving them security and values on which to build the life which awaits them. How many of a three-year old's little questions and reflections are never raised in the boisterous atmosphere of a day nursery? Is it really so hard to understand that there are other values in life than those which consist in doing something useful, something which can be measured in money?[28]

In 1966 a Swedish government survey reported that over half of all women who had children under ten years of age and who were working less than fifteen hours a week did *not* wish to take full-time employment even if child care facilities could be provided. The debate over the desirability of child care outside the nuclear family is a familiar one in our own society, and the same issues have been raised in Sweden where the communal alternative is a more available and socially approved reality.

Despite official government encouragement and a climate of opinion that is favorable to innovative life styles, sex role change has come slowly to Sweden. In 1970 about 55 percent of married women were working outside the home, with the percentage expected to rise to 67 percent in 1985, figures comparable to our own and to those of other industrialized nations. The area most resistant to change has been that of occupational selection. Despite massive government attempts to emphasize the openness of all careers to both men and women by revamping the educational curriculum to counter stereotypes, career choices have remained amazingly traditional. The figures from one regional vocational program tell the tale: the courses chosen by boys ranged from auto mechanics to computer work to draftsmanship, while those chosen by girls were concentrated in teaching, nursing, hairdressing, and clerical work. Women also chose fields requiring less education. In 1966, 37 percent of B.A. graduates were women, but only 8 percent of the candidates for the doctoral degree. Medicine is male, nursing female; the pattern is a familiar one.

The relatively low percentage of women working can be a carryover of traditional attitudes coupled with the largely agrarian background of much of Sweden. The sex-bound nature of the occupational profile can also be attributed to the tenacity of older attitudes and the practical difficulties of switching to another field in mid-career. However, the sex-linked nature of *new* career choices among the young offers no such ready explanation. At least since the early 1960s, young Swedes have grown up in a climate that specifically favors sex role innovation and even provides the practical arrangements (child care facilities and parental allowances) that make such change possible. Yet the sex-bound career profile of this generation looks little different from that of preceding cohort groups. This is especially true in the case of males switching over to so-called "female" jobs, such as nursery school teacher. The old shibboleths against such sex crossover still remain strong and may be transmitted through parental attitudes and peer pressures, which can exert far stronger and longer-lasting effects than official ideology. The power structure of Sweden is basically male and the role models available to young Swedes, both in their families and communities, are for the most part sex-typed in traditional life styles. The special cases that are paraded out for the young in sex role training programs and in rewritten textbooks cannot have the long-term effect that parental attitudes and models do, working together with traditional community mores and sex role structures. Or if such official programs do have an effect, it will probably be fully discernible only after several generations have been through it and social conditions and attitudes have had a real opportunity to change. Women are still choosing "until" jobs, even when they choose to have few or no children, while men are vocationally oriented even when the economic support function has been deemphasized. Leijon comments:

According to a study published in 1963, boys try to get into senior high school education (gymnasium) because it is required for the professions they have chosen. If girls make the same choice, it is because they enjoy being at school, because they do not know what they will become in adult life, or to get a good general education. Many girls refrain from applying because they have no professional ambitions, because senior high school is not required for the occupation of their choice, or because they do not think it worth a girl's while to pass an exam and become an undergraduate.[29]

When we look at the Swedish case, then, we see several trends emerging. There is a necessary lag between announcement of official policies and changes in behavior. In the critical area of *life plans,* a concept we have been emphasizing throughout, the traditional way is favored because it has more psychological and social support than the innovative path. Since women's work has been accorded low social status (despite official protestations to the contrary), men will be even more reluctant than women to change, since such change necessarly means a reduction in status. While the societal power structure is male, and the top of the occupational structure is male, the tendency is for the system to be a self-perpetuating one, with women in important roles outside their families as the exception rather than the rule. It is not just women who are choosing the old paths, but men as well who are experiencing the same achievement pressures their fathers did (but that their mothers and sisters never did). It will clearly take more than a few years to judge the outcome of the Swedish sex role experiment, but the lines of resistance make clear just how tenaciously anchored the *ancien régime* of sex roles is.

Throughout our discussion of "change through space," then, we have seen innovative efforts bog down at several critical points, and it is these very points that show us the true nature of the sex role system. Traditional attitudes linking men with the public sphere and women with the private, domestic one seem to underlie much of the resistance to sex role change. Such attitudes assume that work outside the home will be a secondary concern for women and thereby encourage the maintenance of two different labor markets, one for men, the other for women. These markets are based on the power structure of the society; hence, the upper-status jobs are predominantly male, while the lower-status positions are defined as female. Many of these lower-status female jobs are closely allied to the domestic role, and are really only paid extensions of it. Women continue to be socialized into wanting these jobs, while societal opinion remains strong against men's taking less than their due: a position in the higher-status male labor market. Complicating the picture and in some sense creating much of it is the issue of child care and household work. Even when women are brought into the labor force, the primary responsibility for children and the home is not shared by men, nor do the norms of the society favor such transfer since the identification of women with the domestic sphere is so great. Women seem for the most part to accept these responsibilities, whether it be because of the residue of a biological imperative, socialization and modeling experiences, or the default of the men in refusing to participate, or (probably) a combination of all of the above. The idea of "natural domains" for the two sexes remains strong and is reflected in the male political and economic structure and the female domestic and child care structure. The results of all the sex role experiments described above can be considered to be the "conditional emancipation" of woman, to use Eva Moberg's apt phrase: increased female employment with little accompanying change in the male sex role or in the primary identification of women with the domestic sphere. Acceptable child care alternatives to that of the nuclear family seem to be a stumbling block as well, given the very strong cultural committment to maternal child care. In none of the societies described has women overcome her status as the "second sex" or have underlying traditional attitudes about the proper "place" of men and women been fundamentally altered in action and lifestyles, rather than merely in words or ideology. Men have basically stood firm in their position in the sex role system, as have women in their primary identification with the home and child care; without fundamental changes in these spheres, no real revolution in the sex role system can occur. And we can conclude that in the societies we have studied, it hasn't. What the future holds is uncertain, although it is likely that if societal pressures continue, some evolution in sex roles will occur, though probably over generations rather than years. Whether complete revolution, that is, a life plan and a society not patterned on sex at all, is likely is much more debatable. The inputs of the biological, psychological, and sociological maintenance systems have created a set of values and societal structures that are very resistant to change.

We turn now to consider the history of feminist and sex role thought, particularly the American case, as a source of insight into possible future trends in sex role change and stability.

CHANGE THROUGH TIME

The modern feminist movement is commonly thought to have begun in 1792 with the publication of Mary Wollstonecraft's *A Vindication of the Rights of Woman.* This remarkable book grew out of the climate of the Enlightenment that captured Europe in the years surrounding the French Revolution. Common to her book and to later works, such as John Stuart Mill's *On the Subjection of Women,* is an appeal to reason as the means to social reform. Alice Rossi comments on this line of argument:

> Throughout their writings there is an expectation that "free enquiry" is the simple path to truth which, once arrived at, must absolutely persuade and easily translate into social reality. Their enthusiastic, if naive, belief in education as the cure-all for human ignorance and corruption flowed directly from their confidence in the human cognitive faculty. . . . What "reasonable" person, they seemed to ask, could fail to be convinced by their analysis of the position of women, and once convinced, what "reasonable" man could fail to encourage women to cultivate a wide range of options and skills?[30]

Despite their social psychological and political naivete, such writers as Wollstonecraft and Mill showed considerable insight into the female condition and its sources. Let us listen to Wollstonecraft, recognizing the imprisoning aspects of the "feminine character," even when couched in supposedly positive terms of delicacy and refinement:

> My own sex, I hope, will excuse me, if I treat them like rational creatures, instead of flattering their *fascinating* graces, and viewing them as if they were in a state of perpetual childhood, unable to stand alone. . . . I earnestly wish to point out in what true dignity and human happiness consists – I wish to persuade women to endeavour to acquire strength, both of mind and body, and to convince them that the soft phrases, susceptibility of heart, delicacy of sentiment, and refinement of taste, are almost synonymous with epithets of weakness, and that those beings who are only the objects of pity and that kind of love, which has been termed its sister, will soon become objects of contempt.[31]

Here is Wollstonecraft on the ideal of feminine beauty:

Taught from their infancy that beauty is woman's sceptre, the mind shapes itself to the body, and roaming round its gilt cage, only seeks to adore its prison. Men have various employments and pursuits which engage their attention, and give a character to the opening mind; but women, confined to one, and having their thoughts constantly directed to the most insignificant part of themselves, seldom extend their views beyond the triumph of the hour.[32]

Lastly, Wollstonecraft answers the (still) commonly heard assertion that female intellectual growth and sexual attractiveness to a future marriage mate are incompatible:

If all the faculties of a woman's mind are only to be cultivated as they respect her dependence on man; if, when a husband be obtained, she have arrived at her goal, and meanly proud rests satisfied with such a paltry crown, let her grovel contentedly, scarcely raised by her employments above the animal kingdom; but, if, struggling for the prize of her high calling, she look beyond the present scene, let her cultivate her understanding without stopping to consider what character the husband may have whom she is destined to marry. . . . that she should avoid cultivating her taste, lest her husband should occasionally shock it, is quitting a substance for a shadow.[33]

For her views Wollstonecraft was termed "a hyena in petticoats" by Horace Walpole, and her work was given no recognition in her lifetime or even well beyond it. Although her ideas must surely have shocked the countenance of her time, in reality they did not threaten the bedrock of the sex role system. She did not urge any fundamental alterations in the nature of the family or of male domination in the social structure. In fact, Wollstonecraft uses the female maternal role as an argument for the importance of educating women more extensively:

As the care of children in their infancy is one of the grand duties annexed to the female character by nature, this duty would afford many forcible arguments for strengthening the female understanding, if it were properly considered.[34]

She mentions the possibility of female employment outside the home almost in passing, as a natural outgrowth of education and enlightenment, and as a way out of choosing an unwanted marriage merely because of the necessity for economic support. The gist of her arguments is that rational knowledge and understanding on the part of both men and women will pave the way toward a transformation of woman's state.

John Stuart Mill basically followed the same line of reasoning, that is, an appeal to reason as the road to reform. He also had incisive views of the female condition and combined these with a more direct and scathing denunciation of the marriage and family system. His book came out in 1869, seventy-seven years after Wollstonecraft's, and the difference in tone presages a change toward activism in feminist thinking (the Seneca Falls convention took place in 1848, twenty-one years before Mill's work appeared). Here is a sampling of Mill's thought:

> All causes, social and natural, combine to make it unlikely that women should be collectively rebellious to the power of men. They are so far in a position different from all other subject classes, that their masters require something more from them than actual service. Men do not want solely the obedience of women, they want their sentiments. All men, except the most brutish, desire to have, in the woman most nearly connected with them, not a forced slave but a willing one, not a slave merely, but a favourite. . . . The masters of women wanted more than simple obedience, and they turned the whole force of education to effect their purpose. All women are brought up from the very earliest years in the belief that their ideal of character is the very opposite to that of men; not self-will, and government by self-control, but submission, and yielding to the control of others. All the moralities tell them that it is the duty of women, and all the current sentimentalities that it is their nature to live for others; to make complete abnegation of themselves, and to have no life but in their affections.[35]

Mill subscribed to what we would now term a patriarchal analysis of female oppression, stressing the position of women as a subject class under the power of men. Through the appeal to reason and political reform (notably suffrage), opportunity was sure to follow. Such freedom would benefit humanity, since it would free the heretofore bound minds and spirits of half the human race for higher goals and contributions to civilization. And as a matter of simple justice, the freeing of women was a positive step forward for mankind. For its time, the tenor of Mill's work was revolutionary and gained him much notoriety and disapproval, but the practical repercussions (except in the minds of other feminists) were few. *The Subjection of Women* did definitively state the view that the female character was a product of social conditioning rather than of biological or divine imperative:

> I deny that any one knows, or can know, the nature of the two sexes, as long as they have only been seen in their present relation to one another. . . . What is now called the nature of women is an eminently artificial thing—the result of forced repression in some directions, unnatural stimulation in others. It may be asserted without scruple, that no other class of dependents have had their character so entirely distorted from its natural proportions by their relations with their masters. . . .[36]

Such a view was to appear and reappear time and again as a constant thread through the tapestry of feminism in the next century. The denial of biology and theology in favor of social forces was a necessary first step toward the development of political programs premised on structural reform of sex roles. However, as we shall see, in the twentieth century the pendulum has swung back and forth between acceptance of separate preordained social destinies for women and men, and programs aimed at cutting away the underpinnings of the very sex role system itself. At present we stand somewhere in the middle, with considerable uncertainty among feminists about how to proceed, or in some cases, about *whether* to proceed. For the telling of that tale we turn to the history of feminism in America.

HISTORY OF AMERICAN FEMINISM

As I write this section on the history of American feminism I have before me two small buttons, one saying "Votes for Women," the other, "Opposed to Woman Suffrage." I bought them at a church bazaar a few years ago, no doubt the overflow from someone's attic. The buttons are still in good condition, reminding one of the recency of the suffrage battle (the Nineteenth Amendment was not ratified until 1920), and perhaps of some of its fervor. But it would be a mistaken notion to think that nineteenth- and early twentieth-century American feminism was exclusively aimed at the relatively modest goal of the ballot. In fact, almost all the radical women's rights proposals we hear today (and more) were first heard over a century ago. The tale of compromise and settlement upon the vote as an immediately attainable goal is the history of American feminism.

As Alice Rossi[37] points out, early feminists like Wollstonecraft and Mill were solitary thinkers, not political figures or leaders of social movements. From the very beginning, American feminism was a group endeavor, with programs aimed at social change. The inception of the women's movement in this country was almost an accident, a byproduct of perhaps the best-known social campaign of the last century, abolition of slavery. As Papachristou aptly puts it, "It was, strangely enough, out of her concern for others that the American woman found concern for herself."[38] Since the abolition of slavery was defined as a moral issue, it fell well within women's accepted domain of concern, so that from the first women were actively involved in antislavery societies. It was this experience in collective female endeavor that probably laid the structural groundwork for women's later work in their own behalf. Likewise, participation in other moral reform groups, such as the Women's Christian Temperance Union, led to this sense of joint purpose, and opposition from the male power structure often forced the groups to recognize the inferior public position of women and to strive to do something about it. In the case of slavery, many proslavery forces and the press used the issue of female participation in public life as a tool of ridicule against the women activists. The Grimké sisters, Angelina and Sarah, spoke out in antislavery rallies in the 1830s, incurring the wrath of the churches, public officials, and the press. For example, the *Pittsburgh Manufacturer* wrote of Angelina Grimké's appearance before the Massachusetts legislature: "Miss Grimke is very likely in search of a lawful protector who will take her for 'better or worse' for life, and she has thus made a bold dash among the yankee-law-makers."[39] In the Grimkés' replies to such attacks we see the germination of a feminist consciousness:

> All I ask of our brethren is, that they take their feet from off our necks, and permit us to stand upright on that ground which God designed us to occupy. If he has not given us the rights which have, as I conceive, been wrested from us, we shall soon give evidence of our inferiority, and shrink back into that obscurity, which the high souled magnanimity of men has assigned us as our appropriate sphere. . . . Ah! how many of my sex feel in the dominion, thus unrighteously exercised over them, under the general appellation of *protection,* that what they have leaned upon has proved a broken reed at best, and oft a spear.[40]

These words, fighting sentiments couched in none too gentle sarcasm, were written in 1837. The general atmosphere of reform and the requirements of frontier and near-frontier life made for a somewhat freer conception of womanhood in America than on the other side of the Atlantic. Historically, few women were far removed from the frontier past when women's work was a necessity and the leisure-class housewife almost an unknown entity. Although there was some carryover from European ideals of womanhood that colored images of "woman's place," there was clearly room for something else as well that thrived on the freer American climate, unbound by centuries of tradition. Tocqueville in his *Democracy in America,* written after a trip here in the 1830s, noted the difference:

> An American girl scarcely ever displays that virginal softness in the midst of young desires, or that innocent and ingenuous grace, which usually attend the European woman in the transition from girlhood to youth. It is rare than an American woman, at any age, displays childish timidity or ignorance. . . . I have been frequently surprised, and almost frightened, at the singular address and happy boldness with which young women in America contrive to manage their thoughts and their language. . . . It is easy, indeed, to perceive that, even amidst the independence of early youth, an American woman is always mistress of herself. . . .[41]

The first organized group aimed exclusively at woman's rights met at Seneca Falls, New York on July 19 through 20, 1848. The First Woman's Rights Convention took place in the Wesleyan Chapel, a site now occupied by a laundromat[42] (used almost exclusively by women doing wash)—somehow an apt comment on the rocky course of feminism in America. Three hundred people (almost all women) attended the convention, called by five women, Lucretia Mott, Martha C. Wright, Jane Hunt, Elizabeth Cady Stanton, and Mary Ann McClintock. Out of that meeting came a Declaration of Sentiments (based on the Declaration of Independence) and several resolutions. Excerpts from these documents give the flavor of the sentiments:

> The history of mankind is a history of repeated injuries and usurpations on the part of man toward woman, having in direct object the establishment of an absolute tyranny over her. . . . He has never permitted her to exercise her inalienable right to the elective franchise. . . . He has made her, if married, in the eyes of the law, civilly dead. . . . He has monopolized nearly all the profitable employments, and from those she is permitted to follow, she receives but a scanty remuneration. . . . He has denied her the facilities for obtaining a thorough education, all

colleges being closed against her. . . . He has endeavored, in every way that he could, to destory her confidence in her own powers, to lessen her self-respect, and to make her willing to lead a dependent and abject life. . . .[43]

Clearly, the tenor of the remarks goes far beyond the call for the vote, although that was perhaps the most clear-cut of the resolutions and the one that led to the simplest solution; hence, eventually, all the other demands became condensed into this one as the most practically realizable goal.

Many critics, even in that early period, realized what extensive changes in the sex role system were being called for, and they rallied against this threat, with the vote as a symbol. An Albany paper perceptively wrote of the implications of the Seneca Falls resolutions:

Now, it requires no argument to prove that this is all wrong. Every true hearted female will instantly feel that this is unwomanly, and that to be practically carried out, the males must change their position in society to the same extent in an opposite direction, in order to enable them to discharge an equal share of the domestic duties which now appertain to females, and which must be neglected, to a great extent, if women are allowed to exercise all the "rights" that are claimed by these Convention-holders. Society would have to be radically remodelled in order to accommodate itself to so great a change in the most vital part of the compact of the social relations in life; and the order of things established at the creation of mankind, and continued six thousand years, would be completely broken up.[44]

Hence, the discussions that took place over suffrage and women's rights in the ensuing years had the flavor that similar debates have today, that of fear over the threat to the stability of the traditional sex role system.

Additional issues became added to the suffrage ferment, issues of equality in marriage, employment, dress, and freedom of movement and public involvement. Several journals appeared devoted to questions of women's rights, and articles on these wider issues were admixed with the central one of suffrage. Views on women's right to initiate divorce, particularly against drunken husbands, were aired, as well as the more revolutionary doctrine of compensation for household work. For example:

All this talk about the indissoluble tie and the sacredness of marriage, irrespective of the character and habits of the husband, is for its effect on woman. She never could have been held the pliant tool she is today but for the subjugation of her religious nature to the idea that in whatever condition she found herself as man's subject, that condition was ordained of Heaven; whether burning on the funeral pile of her husband in India, or suffering the slower torture of bearing children every year in America to drunkards, diseased, licentious men, at the expense of her life and health and the enfeebling of both the mind and body of her progeny.[45]

Women have borne in silence this position of paupers. . . . It is time that all this was changed—that a woman should become not only in name, but in fact, the equal partner of her husband in the money which he amasses. . . .[46]

Coming at the same time was increased female agitation for education and entry into the professions. The case of Elizabeth Blackwell, who became the first female physician in America, is well known. She linked her cause to that of the general advancement of women, and perceptively noted the effect such initial career entries would have on others:

For what is done or learned by one class of women becomes, by virtue of their common womanhood, the property of all women. It tells upon their thought and action, and modifies their relations to other spheres of life, in a way that the accomplishment of the same work by men would not do.[47]

Although it would be inaccurate to say that a clamor for women's rights was sweeping the nation during the latter part of the nineteenth century, there was an increasing sensitivity among women, especially those of the upper classes and those who were organized into women's groups, to their own welfare and to "women's issues." Colleges for women were established, and women became increasingly self-conscious about their condition. In the 1850s a movement began for dress reform that was symptomatic of rebellion against the restrictive female garments that were indicative of women's social condition. The so-called "Bloomer" outfit with its pantaloons and tunic—and the furor it occasioned—were but a trivial instance of a much more deeply rooted dissatisfaction. Outcries were heard against the double standard of morality that judged women much more harshly, and hence restricted their lives more radically. Such sentiments chipped away at the pedestal of femininity that had been so carefully erected over the generations to cloak restriction with a covering of divine purpose and higher calling than men.

Much of the criticism of the women's rights movement was phrased in terms of the immorality of the reforms, which critics saw as aimed at the destruction of the home and of the sexual innocence of women. Given the Victorian tenor of the time, the charge of advocating "free love"—sex outside of marriage—was a very serious one indeed, and feminists often took pains to couch their reforms in moralistic terms of the ultimate preservation of the family and the elevation of womanhood. They sought to reassure the sexually anxious public that their reforms, particularly suffrage, would not violate the traditional values of feminine purity and the sanctity of the family. In fact, many of the suffragists argued that giving the vote to women would permit them to express their views on issues peculiarly suited to women, such as the family, child welfare, etc. The National Woman Suffrage Association in 1871 maintained that female suffrage would ensure "stability for the marriage relation, stability for the home and stability for our representative form of government."[48]

However, a few feminist activists refused to live within the limits of their times and drew notoriety onto themselves and, by implication, endangered the more modest goals of the women's rights movement. The most prominent of these was Victoria Woodhull, who shocked the nation by her views on free love and her accusation of adultery against Henry Ward Beecher, a prominent preacher. Her views would probably cause no great stir today but had an electric effect in the 1870s and for a while endangered the respectibility and hence the success of the suffrage movement. Here is Ms. Woodhull:

> Yes, I am a Free Lover. I have an inalienable, constitutional and natural right to love whom I may, to love as long or as short a period as I can; to change that love every day if I please, and with that right neither you nor any law you can frame have any right to interfere.[49]

We have so far deemphasized the fight for the vote itself in favor of a more general consideration of the movement for women's rights in the nineteenth and early twentieth centuries. But since it is the vote that now is the only tangible evidence of that movement's existence, it behooves us to examine the path of progress toward that end. Early attempts to include women in the post–Civil War amendments giving blacks the vote and other civil rights failed and led to the activities of two suffrage groups (which were sometimes rivals), the American Woman Suffrage Association and the National Woman Suffrage Association. The latter group was the more radical of the two and aimed at broader goals than the vote. It also took to more belligerent tactics than the American and was less hospitable to men's joining its ranks. The National agitated over issues of wages for women and marriage laws, while the American stuck to the more politically respectable aim of the vote, often couching its arguments in terms of women's special sensitivity and talents in the area of social reform as a positive addition to the electorate. In 1890 the two groups united in the form of the National American Suffrage Association. However, this group still evoked considerable public antipathy, and it was only the prosuffrage position of the morally acceptable Women's Christian Temperance Union that gave impetus to the cause. Although the WCTU began as a crusade against liquor, it quickly realized that it needed political clout to achieve its ends. This realization, combined with the growing liberalism of the times (in part created by the previous half-century of femininst agitation) led to the call for "equal franchise, where the vote of woman joined to that of men can alone give stability to Temperance legislation."[50] As Papachristou put it, "The WCTU offered American women a gentle and respectable transition from the home to a larger world as it involved them in stimulating and provocative public issues."[51] Frances Willard, the head of the WCTU for many years, gradually developed a more encompassing view of women's place in public life:

> The W.C.T.U. is doing no work more important than that of reconstructing the ideal of womanhood. . . . Woman is becoming what God meant her to be and Christ's gospel necessitates her being, the companion and counsellor, not the incumbrance and toy, of man. . . . The world has never yet known half the amplitude of character and life to which men will attain when they and women live in the same world.[52]

Gradually through state and national activism, the idea of female suffrage became more acceptable, although not without considerable acrimonious debate. At the same time that the suffrage debate was going on, numerous other reforms were being introduced into American life. Women were entering the labor force, and ultimately the labor union movement. Other women were involved in club work, sometimes aimed at improving the lot of women and children (particularly those in the labor force) and of the immigrant groups that were spilling into America at that time. A small number of women, notably leaders like Emma Goldman, were gaining prominence in the socialist movement. Black women were organizing to protest their condition. A definite strain of radicalism was being felt in the feminist movement. Charlotte Perkins Gilman wrote *Women and Economics* in 1898 and called for the end of the sexual division of labor and women's economic dependence. For her time, she posed a devastating critique of the sacred notions of the family and female service to it:

> As a natural consequence of our division of labor on sex lines, giving to woman the home and to man the world in which to work, we have come to have a dense prejudice in favor of the essential womanliness of the home duties, as opposed to the essential manliness of every other kind of work. We have assumed that the preparation and serving of food and the removal of dirt, the nutritive and excretive processes of the family, are feminine functions; and we have assumed that these processes must go on in what we call the home, which is the external expression of the family. . . . Is it not time that the way to a man's heart through his stomach should be relinquished for some higher avenue? . . . We need a new picture of our overworked blind god,—fat, greasy, pampered with sweetmeats by the poor worshippers long forced to pay their devotion through such degraded means.[53]

Gilman went on to suggest the communalization of domestic and child care functions, with living complexes provided with common kitchens and day care facilities, permitting the unimpeded entry of both women and men into the labor force. Believeing that the sexual division of labor and the economic dependence of women were keys to the "woman problem," Gilman did not hesitate to put forth her solutions, ideas that we would still consider radical today, nearly eighty years after their inception.

Given the background of such revolutionary ideas, many of which shocked women's rights groups themselves as well as the considerable armies of antifeminists, the suggestion of the vote for women seemed like a tame alternative. Suffragist groups were stepping up their campaigns, buoyed by victories in the states and the granting of female suffrage abroad. The activist National Woman's Party did its part, and both local and national agitation, including such tactics as picketing the White House, increased. The climate of the Progressive Era and World War I, fought to preserve democracy, made ideological opposition to woman suffrage harder to justify. Finally, in 1918, Woodrow Wilson was convinced that his original plan for state-by-state adoption of female suffrage was unworkable, and he came out in support of the Nineteenth Amendment, which was ultimately ratified by the states on August 26, 1920. The climax of the long campaign, and in many ways the end of feminism for another half-century, was embodied in a thirty-nine-word statement:

> The right of citizens of the United States to vote shall not be denied or abridged by the United States or by any State on account of sex. Congress shall have power to enforce this Article by appropriate legislation.

The tale of the ensuing fifty years is one of limited progress for women and, indeed, of considerable retrenchment in terms of the restoration of domesticity as the life goal for women. Contrary to expectations, the winning of the vote for women was the end, not the beginning, of feminist reform in this country. Part of the reason may simply have been that after suffrage, no clear-cut goal existed as a rallying point and an aim for the movement. A campaign was mounted in the 1920s by the National Woman's Party to put through the Equal Rights Amendment—the same amendment that is being debated at the present time. Many of the same issues were raised then, including the advisability of protective labor laws for women; and, of course as we know, the amendment was never passed. The Equal Rights Amendment campaign and others in the twenties attracted a fraction of the support that the suffrage movement had. The conservative tenor of the times after the war and hysteria over socialism led to a growing disfavor with social reform of any kind. In addition, there simply was no cadre of women available to push for such demands. Some change had overtaken the new generation of women. As Chafe notes:

> 'Feminism has become a term of opprobrium (for the young),' Dorothy Dunbar Bromley noted in 1927. 'The word suggests either the old school . . . who wore flat heels and had very little feminine charm, or the current species who antagonize men with their constant clamor about maiden names.' Neither type had any appeal for the college girl. She enjoyed the benefits the feminists had won, but refused to consider their cause her own. 'We're not out to benefit society . . . or to make industry safe,' one student commented. 'We're not going to suffer over how the other half lives.' As a League of Women Voters official la-

mented, the woman's movement had ceased to hold any attraction for the 'juniors.'[54]

The vote itself did not hold any magical power; fewer women than men even bothered to exercise their franchise, and those who did voted much like their husbands, leading to little net change in the ballot. Earlier suffragist hopes of a bloc of women's votes that would make politicians stand up and listen fizzled. Very few women took part in political activities, and the vote ultimately faded into the background as a necessary but not crucial step in the feminist march toward equality.

The "flapper era" of the twenties brought a freer public image of women, implying a new openness in sexuality and expression that could have never emerged in the preceding century. However, for most women, life remained much the same as before. No great increase in female employment was noted (the real upswing had occurred at the turn of the century), and the jobs in which women were employed were squarely within "woman's place." Chafe found that in the 1920s, "over 40 per cent of all women in manufacturing were employed in textile mills or as apparel operatives, and more than 75 per cent of female professionals were either teachers or nurses."[55] Little opportunity for advancement was given, and then only within sex-stratified lines. A common belief existed that women worked only for "pocket money," and hence should not be given equal consideration with men who had to support families.

Higher education for women was becoming more common, but there was a decided turn toward a "women's curriculum" and away from the earlier model of creating women's colleges that would be the equals of the best male institutions. Education for domesticity and motherhood rather than careers became the norm, and formal and informal pressures against women professionals began to take their toll. The return to the traditional model was almost complete, with some changes, of course. The fight for birth control, perhaps one of the most important steps in the feminist movement, was coming to a successful conclusion after a difficult battle against conservative forces outraged at the very mention of sex and/or any interference with "natural processes." Margaret Sanger, the leader of the birth control movement, saw it in a broader context, one of female self-determination:

> The problem of birth control has arisen directly from the effort of the feminine spirit to free itself from bondage. . . . it is woman's duty as well as her privilege to lay hold of the means of freedom. Whatever men may do, she cannot escape the respon-

> sibility. For ages, she has been deprived of the opportunity to meet this obligation. She is now emerging from her helplessness. Even as no one can share the suffering of the overburdended mother, so no one can do this work for her. Others may help, but she and she alone can free herself.[56]

The availability and social acceptability of birth control made a tremendous impact on the potential separation of women from the home, but it did not complete the job. Women were still psychologically wedded to their place and men stayed put in theirs. The militarism of World War I and the consequent sex-separation of male fighter and female home-fire-stoker might have accelerated the pace of return to the traditional, as occurred in the period after World War II with its tremendous baby boom and institutionalization of "feminine mystique" values. Perhaps any society can absorb just so much change at once and the suffrage seemed change enough. Even at the height of nineteenth- and early twentieth-century feminism, only a small minority of women (and a smaller minority of men) were involved. Most rationalized their concern with the vote as a way of furthering woman's unique contribution to society, through her intimate association with family and child concerns. Very, very few thought in the radical terms of women like Charlotte Perkins Gilman, and fewer still could forge a new way for themselves in the face of almost complete societal and structural opposition.

The Great Depression of the 1930s worsened the economic position of women even more so than that of men. World War II brought a temporary widening of opportunities, as through necessity women were permitted to take jobs that had previously been assigned to men. The image of "Rosie the Riveter" adorned many a war poster and magazine cover, but after the war it was forgotten. Some women did remain in the labor force, but usually in the "women's jobs" left after the returning veterans reclaimed their old positions. Many more women, however, were contributing to the "baby boom" of the late forties, living the traditional life of domesticity and child care. The growth of the suburbs during the postwar era further dispersed women, separating them from job opportunities in the cities and creating a subculture of suburbanism that encouraged the identification of femininity exclusively with home and family life. Books appeared, such as Lundberg and Farnham's *Modern Woman: The Lost Sex,*[57] which supported the traditional feminine role and cast aspersions on career goals for women as indicative of "masculinity strivings." The circle from feminism was complete and was closing around the life of the American woman.

Betty Friedan, in her book *The Feminine Mystique,*[58] began in 1963 a new wave of feminine awareness that culminated in the current women's movement. Earlier books such as Simone de Beauvoir's *The Second Sex*[59] had appeared in the 1950s, but these were not addressed to a mass audience and were written too abstractly to stimulate action or debate. Friedan's book, on the other hand, spoke directly to the condition of the American woman, or rather, her social conditioning by advertising and schooling. She argued persuasively that this "mystique," as she called it, of glorification of domestic and family values above self-fulfillment, spelled dissatisfaction for women. The book sparked an immediate debate in the media and led to retorts, such as that by Phyllis McGinley, extolling the virtues of housewifery. The issues were readily understandable to American women because they dealt with their everyday environment and concerns and seemed to identify a common fate, one that was not satisfactory to many women. The private complaints of many became transformed into a public issue. To put it in more abstract terms, there was a "consciousness" of the shared experience of women and the widespread problems such a fate imposed. Instead of dismissing their private gripes, many unhappy or uncertain women could identify with a sex role fate. This identification was to provide the impetus for the new feminism of the sixties and seventies. In addition, the general liberal climate of the times, which encouraged rights for blacks and other minorities, also contributed to the shaping of the movement.

Friedan's analysis involved a discussion of what she called the "sexual sell," in which female guilt over nonfulfillment of the domestic role was used as a means to sell consumer products. The image of womanhood projected in the media, particularly in advertising and women's magazines, was one of unalloyed delight in the traditional feminine role, an image contradicted by the feelings of many women. A leisure class of housewives had emerged as the ideal, and even with the increasing number of women in the labor force, this ideal was exerting a psychological hold over the life plans of young women. Education for women largely aimed at filling time before marriage or at developing facility at cocktail party repartee; seldom did it help women develop professional aims. Women were encouraged by Freudian-directed educators and therapists as well as consumer-oriented advertisers to live through their husbands and children and their home. All of these values, which Friedan collectively termed the "feminine mystique," served to perpetuate the traditional sex role system. She called for a "new life plan for women" to save the "forfeited self," a life plan that would allow women to advance in professional careers and develop confidence in their own abilities and aims, rather than seek satisfaction through those of others. Friedan ended her book on much the same note that feminists had 150 years before:

> Who knows what women can be when they are finally free to become themselves? Who knows what women's intelligence will contribute when it can be nourished without denying love? Who knows of the possibilities of love when men and women share not only children, home, and garden, not only the fulfillment of their biological roles, but the responsibilities and passions of the work that creates the human future and the full human knowledge of who they are? It has barely begun, the search of women for themselves. But the time is at hand when the voices of the feminine mystique can no longer drown out the inner voice that is driving women on to become complete.[60]

In the years following Friedan's book, while the social ferment of the sixties was developing, two distinctive ideologies for sex role change came into prominence. The first, which we shall call the *reformist* ideology, argues for the equality of women within the existing sex role system. Proponents of this view advocate legislative change, increased opportunity for women to enter management and the professions, and an end to discriminatory practices against women. Such changes would "elevate" the position of women to that of men without requiring radical change in the organization of the family or of society. Guettel terms this the "liberal tradition," "a moralistically humanitarian and egalitarian philosophy of social improvement through the reeducation of psychological attitudes."[61] Such a philosophy is akin to that of the early Enlightenment feminists, such as Wollstonecraft and Mill, with an added note of political and social practicality. Consonant with this position would be attempts to remove sex role stereotypes from textbooks, television, and advertising, and to encourage a less sexually-demarcated upbringing of children. Men are not primarily seen as enemies, and in fact are welcomed in the fight for feminism, and the basic nature of heterosexual relationships and family life is not questioned. Child care facilities and the sharing of housework might be advocated, but as an adjunct to the present social system, not a replacement for it. The tacit acceptance of the identification of home and family life with the female is made, although societal arrangements (such as day care) to lessen the burden for the woman are encouraged. Little is said about male roles, except for the hope that more latitude in female roles will naturally lead to a greater sharing of human experiences and feelings, and consequently to a less rigid definition of masculinity as well. It would probably be safe to say that most women who identify themselves with the feminist movement today subscribe in one way or another to this reformist view.

The second view, the *revolutionary* ideology, offers a much more devastating critique of the sex role system, and, usually, of the accompanying societal structure as well. Many of the proponents of this view support the Marxist analysis of society and its consequences for the sex role system. They often speak of women explicitly as an oppressed class, and identify men and the patriarchal family system, as the oppressors. Yates, who terms this view the women's liberationist paradigm, describes it as follows:

> It is a conflict and confrontation model, whereby women collectively assert their own importance, call for solidarity among women, and assert a new politics whereby the male definition of woman as sex object is turned into a political tool for forging a new political order out of the special experiences of women. It is a pattern of women-over-against-men, in which new norms will come from women.[62]

Some of the founders of the revolutionary movements carried over their views from their earlier experiences in New Left movements of the sixties, in which they perceived the sexist attitudes of their male colleagues and seceded to form groups based on their own grievances as women. They developed a radical critique of the sex role system, sometimes rejecting heterosexuality itself as a paradigm for male oppression, but always aiming at the present structure of marriage and the family and emphasizing the sisterhood of women. Much attention has been focused on eliminating "feminine mystique" values, such as those of personal adornment, as an impediment to self-realization. Such well-publicized incidents as the picketing of the Miss America pageant reveal this attitude, although they attracted publicity far out of proportion to their true significance. The ultimate aim of the revolutionary movement is the overthrow of the sex role system itself. Shulamith Firestone in *The Dialectic of Sex* expresses this view, contrasting it to the reformist position:

> And just as the end goal of socialist revolution was not only the elimination of the economic class privilege but of the economic class distinction itself, so that the end goal of feminist revolution must be, unlike the first feminist movement, not just the elimination of male privilege but of the sex distinction itself: genital differences between human beings would no longer matter cul-

> turally. . . . The reproduction of the species by one sex for the benefit of both would be replaced by (at least the option of) artificial reproduction: children would be born to both sexes equally, or independently of either, however one chooses to look at it; the dependence of the child on the mother (and vice versa) would give way to a greatly shortened dependence on a small group of others in general, and any remaining inferiority to adults in physical strength would be compensated for culturally. The division of labour would be ended by the elimination of labour altogether (cybernation). The tyranny of the biological family would be broken.[63]

Practical ways of implementing such policies usually feature some sort of Marxist program of production, although as we have seen, the history of such attempts in socialist countries up to now has not been extremely successful. Other radical groups have given up on the possibility for immediate change and instead advocate complete separation from men as the only practical solution.

The current climate is one of increasing awareness of the oppressive features of the sex role structure, with reformist legislation being a favored means of change. Other changes are also notable; for example, those in public life are less likely to cast aspersions on women with impunity, a change that could be compared to that occurring with the treatment of blacks and other minority groups. Educational institutions such as Yale and West Point have been opened to women—unthinkable steps a generation ago. Feminist writings are becoming a mass product with the appearance of publications like *Ms.* magazine. Even traditional women's magazines have broadened their scope beyond the confines of the family and domestic life. Token women in unusual occupations (for women) such as telephone line repair work are touted in the press. Even the language we use is undergoing change: the appearance of words like "chairperson," for example, to replace "chairman." We now have the limited use of the term "Ms." to signify the social separation of the female from her marital status. Credit laws have been changed to permit realization of female financial independence. And we could go on with other examples.

A new dimension to the sex role debate has been added with the critique of the male role and its emphasis on competitive achievement and the holding back of emotions. The Gay Liberation movement has been instrumental in broadening the permissible definitions of masculinity and in sparking the debate on the meaning of masculinity. Clearly, any deep change in the sex role system must imply a redefinition of masculinity as well as of femininity.

Men's consciousness raising groups have arisen alongside women's to discuss the issues and explore alternatives.

There has been a backlash against the new feminism, notably the appearance of "Total Woman" and "Fascinating Womanhood" courses and books, which stress the joys of the traditional feminine role. Acting as a brake on mass enthusiasm as well has been the radical image given to even the moderate proponents of change by the more revolutionary parts of the movement, for example, lesbian activists. At the present time, it would be safe to say that most Americans, female and male, are living their lives within the structure of the traditional sex role system but are being exposed to an environment encouraging questioning of that system and postulation of alternatives to it. As we have seen in this chapter and the preceding ones, however, the sex role system is solidly entrenched in our biological, psychological, and social systems, and the pace of change will probably be slower than any of us can imagine. But the possibility is there, and the lesson of history is one of continuous change. Perhaps it will ultimately extend to that most resistant of all domains, the sex role system, and expand and enrich the future of both sexes.

SUGGESTED READINGS

Ester Boserup. *Woman's Role in Economic Development*. New York: St. Martin's Press, 1970 (p). How women fare in Third World countries.

Bernice A. Carroll, ed. *Liberating Women's History: Theoretical and Critical Essays.* Urbana: University of Illinois Press, 1976 (p). Anthology of studies of women's history around the world.

William H. Chafe. *The American Woman: Her Changing Social, Economic and Political Role, 1920–1970*. New York: Oxford University Press, 1974 (p). Excellent book on the transitional period for American women between the two feminist movements.

Edmund Dahlström, ed. *The Changing Roles of Men and Women*. Boston: Beacon Press, 1971 (p). Sex roles in Sweden studied from a number of social science perspectives.

Delia Davin. *Woman Work: Women and the Party in Revolutionary China*. New York: Oxford University Press, 1976. Authoritative account of the changes in women's lives under the Chinese Communist regime.

Betty Friedan. *It Changed My Life: Writings on the Women's Movement*. New York: Random House, 1976. Memoirs of the author of *The Feminine Mystique* and one of the leaders of the new feminist movement.

Linda Gordon. *Woman's Body, Woman's Right: A Social History of Birth Control in America*. New York: Viking Press, 1976. Well-done history about one of the central movements that changed women's lives.

Vivian Gornick and Barbara K. Moran, eds. *Women in Sexist Society: Studies in Power and Powerlessness*. New York: New American Library, 1972 (p). One of the best collections of feminist thought of the 1960s and 1970s.

Charnie Guettel. *Marxism and Feminism*. Toronto: The Women's Press, 1974 (p). Concise presentation of various theorists' views.

Mary Hartman and Lois W. Banner, eds. *Clio's Consciousness Raised: New Perspectives on the History of Women*. New York: Harper & Row, 1974 (p). Good, basic collection.

Judith Hole and Ellen Levine. *Rebirth of Feminism*. New York: Quadrangle Books, 1971 (p). History of the latest movement for change.

Anne Koedt, Ellen Levine, and Anita Rapone, eds. *Radical Feminism*. New York: Quadrangle Books, 1973 (p). Collection of classic articles from feminist publications of the 1960s and early 1970s.

Aileen S. Kraditor. *The Ideas of the Woman Suffrage Movement, 1890–1929*. New York: Doubleday, Anchor Books, 1965 (p). The old tale of radicalism giving way to conservatism. Issues are handled clearly.

Carolyn J. Matthiasson, ed. *Many Sisters: Women in Cross-Cultural Perspective*. New York: The Free Press, 1974. Good collection of studies from both well-known and less well-known societies, illuminating the position of women.

Kate Millett. *Sexual Politics*. New York: Doubleday, 1970 (p). Political relationship between the sexes as seen in psychology, history, and literature. A much talked-about book of the early 1970s.

William L. O'Neill. *Everyone Was Brave: The Rise and Fall of Feminism in America*. New York: Quadrangle Books, 1969 (p). Excellent, readable account of why women got the vote but little else.

Judith Papachristou. *Women Together: A History in Documents of the Women's Movement in the United States*. New York: Knopf, 1976 (p). Fine collection of materials, some well-known, some not, set off by excellent transitional material.

Alice S. Rossi, ed. *The Feminist Papers: From Adams to Beauvoir*. New York: Bantam Books, 1974 (p). Good collection set off by Rossi's excellent introductory material. More emphasis on theory than there is in Papachristou.

Mary P. Ryan. *Womanhood in America: From Colonial Times to the Present*. New York: Franklin Watts, New Viewpoints, 1975 (p). Wide-ranging presentation over a broad time span. A good overview of the subject.

Hilda Scott. *Does Socialism Liberate Women? Experiences from Eastern Europe*. Boston: Beacon Press, 1974 (p). What happened in Eastern Europe when women's roles changed, but the change went only so far. Well-integrated book which also has a very clear presentation of Engels' theory of the family.

Georgene H. Seward and Robert C. Williamson, eds. *Sex Roles in Changing Society*. New York: Random House, 1970. Good anthology of articles on the status of women around the world.

Ruth Sidel. *Women and Child Care in China: A Firsthand Report*. Baltimore: Penguin Books, 1973 (p). The nth book about China, but a good one.

Claire Tomalin. *The Life and Death of Mary Wollstonecraft*. New York: New American Library, 1974 (p). Well-received biography of a woman before her time.

Maggie Tripp, ed. *Woman in the Year 2000*. New York: Arbor House, 1974 (p). Some fanciful but mostly realistic guesses about the future of women.

Mary Wollstonecraft. *A Vindication of the Rights of Woman*. 1792; rpt. New York: Norton, 1967 (p). One of the earliest feminist declarations.

Gayle Graham Yates. *What Women Want: The Ideas of the Movement*. Cambridge, Mass.: Harvard University Press, 1975. History and analysis of the current women's rights movement.

NOTES

INTRODUCTION

1. O'Neill, W. L., *Everyone was Brave: The Rise and Fall of Feminism in America,* Quadrangle, Chicago, 1969, p. 358.

2. Mead, M., *Sex and Temperament,* Morrow, New York, 1935.

3. Fortune, W. F., "Arapesh warfare," *American Anthropologist* 41 (1939): 22–41.

CHAPTER 1. The Biological Maintenance System

1. Lorenz, K., *On Aggression,* Harcourt Brace Jovanovich, New York, 1966.

2. "Women's lib, Amazon style," *Time*, December 27, 1971, p. 54.

3. For an excellent discussion of biological factors and their interrelationship with social factors in sexual differentiation, see Money, J., and Ehrhardt, A. A., *Man and Woman, Boy and Girl,* Johns Hopkins Press, Baltimore, 1972.

4. Klüver, H. and Bucy, P. C., "Preliminary analysis of functions of temporal lobes in monkeys," *Archives of Neurology and Psychiatry* 42 (1939): 979–1000.

5. See Mark, V. H., and Ervin, F. R., *Violence and the Brain,* Harper & Row, New York, 1970.

6. Ibid., pp. 99–108.

7. Raisman, G., and Field, P. M., "Anatomical considerations relevant to the interpretation of neuroendocrine experiments," in Martini and Gonong, eds., *Frontiers in Neuroendocrinology,* Academic Press, New York, 1971, p. 14.

8. Ibid., "Sexual dimorphism in the preoptic area of the rat," *Science* 173 (1971): 731–733.

9. Arnold, M., "Emotion, motivation, and the limbic system," *Annals of the New York Academy of Sciences* 159(3) (1969): 1041–1058.

10. Smythies, J. R., *Brain Mechanisms and Behavior,* Academic Press, New York, 1970, p. 156.

11. See Moyer, K. E., "Sex differences in aggression," in R. C. Friedman, R. M. Richart, R. L. Vande Wiele, eds., *Sex Differences in Behavior,* Wiley, 1974, pp. 335–372.

See also Money and Erhardt, note 3. See also Hart, B. L., "Gonadal androgen and sociosexual behavior of male mammals: A comparative analysis," *Psychological Bulletin* 81(7) (1974): 383–400.

12. Goy, R. W., "Organizing effects of androgen on the behaviour of rhesus monkeys," in R. P. Michael, ed., *Endocrinology and Human Behaviour*, Oxford University Press, London, 1968.

13. See Hook, E. B., "Behavioral implications of the human XYY genotype," *Science* 179 (1973): 139–150; also Meyer-Bahlburg, F. L., "Aggresion, androgens and the XYY syndrome," in R. C. Friedman, R. M. Richart, R. L. Vande Wiele, eds., *Sex Differences in Behavior,* Wiley, New York, 1974, pp. 433–454.

14. Money, J., "Impulse, aggression and sexuality in the XYY syndrome," *St. John's Law Review* 44 (1970): 220–235.

15. Cowie and Kahn, "XYY constitution prepubertal child," *British Medical Journal* (1968): 748–749 (cited by Money, above, pp. 230–231).

16. Hook, p. 147.

17. Sherfey, M. J., *The Nature and Evolution of Female Sexuality,* Random House, New York, 1972.

18. Ibid., p. 144.

19. Degler, C., "What ought to be and what was: Women's sexuality in the nineteenth century," *American Historical Review* (1974): 1467–1490.

20. Bardwick, J. M., *The Psychology of Women: A Study of Biocultural Conflicts,* Harper & Row, New York, 1971.

21. McClelland, D., "Wanted: A new self-image for women," in R. J. Lifton, ed., *The Woman in America*, Beacon Press, Boston, 1965, pp. 173–192.

22. Ibid., p. 185.

23. Eibl-Eibesfeldt, I., "Similarities and differences between cultures in expressive movements," in R. A. Hinde, ed., *Nonverbal Communication,* Cambridge University Press, Cambridge, 1972, pp. 297–312.

24. Kolata, G. B., "Primate behavior: Sex and the dominant male," *Science* 191 (1976): 55–56.

25. Bardwick, J. M., "Psychological conflict and the reproductive system," in Bardwick et al., eds., *Feminine Personality and Conflict,* Brooks Cole, Belmont, California, 1970, pp. 3–30.

26. Douglas, M., *Purity and Danger: An Analysis of Concepts of Pollution and Taboo,* Penguin, Baltimore, 1966, pp. 174–5 (from Meggitt, M., "Male-female relationships in the highlands of New Guinea," in *American Anthropologist,* special publication 2, *New Guinea: The central highlands,* J. B. Watson, ed., 1964).

27. Ernster, V. L., "American menstrual expressions," *Sex Roles* 1(1) (1975): 3–14.

28. See Maddox, H. C., *Menstruation,* Tobey Publishing Co., New Canaan, Connecticut, 1975. Also Persky, H., "Reproductive hormones, moods, and the menstrual cycle," in R. C. Friedman, R. M. Richart, and R. L. Vande Wiele eds., *Sex Differences in Behavior*, Wiley, New York, 1974, pp. 455–476.

29. Dalton, K., *The Menstrual Cycle,* Pantheon, New York, 1969.

30. Parlee, M. B., "The premenstrual syndrome," *Psychological Bulletin* 80 (1973): 454–465. In Cox, S., ed., *Female Psychology: The Emerging Self,* Science Research Associates, Chicago, 1976.

31. From preliminary version of Parlee, above, p. 5.

32. Coppen, A., and Kessel, N., "Menstruation and personality," *British Journal of Psychiatry* 109 (1963): 711–721.

33. Ivey, M. E., and Bardwick, J. M., "Patterns of affective fluctuation in the menstrual cycle," *Psychosomatic Medicine* 30(3) (1968): 336–344. In R. K. Unger and F. L. Denmark, eds., *Woman: Dependent or Independent Variable?* Psychological Dimensions, New York, 1975, pp. 538–552.

34. Ibid., p. 548.

35. Paige, K. E., "Effects of oral contraceptives on affective fluctuations associated with the menstrual cycle," *Psychosomatic Medi-*

cine 33(6) (1971): 515–37. Also in Unger and Denmark, pp. 554–589.

36. Schrader, S. L., Wilcoxon, L. A., and Sherif, C. W., "Daily self-reports on activities, life events, moods, and somatic changes during the menstrual cycle," a paper given at the American Psychological Association Convention, Chicago, August 1975, p. 45.

37. Koeske, R. K., and Koeske, G. F., "An attributional approach to moods and the menstrual cycle," *Journal of Personality and Social Psychology* 31(3) (1975): 473–478.

38. Ibid., pp. 477–478.

39. Koeske, R. K., "'Premenstrual tension' as an explanation of female hostility," paper given at American Psychological Association meeting, Chicago, September 1975.

40. Ramey, E., "Men's monthly cycles," in F. Klagsbrun, ed., *The First Ms. Reader*, Warner, New York, 1973, pp. 174–181.

41. Halberg, V. P. and Hamburger, C., "17-ketosteroid and volume of human urine: Weekly and other changes with low frequency," *Medicine* 47 (1964): 916–925.

42. Doering, C. H., Brodie, H. K. H., Kraemer, H., Becker, H., and Hamburg, D. A., "Plasma testosterone levels and psychologic measures in men over a 2 month period," in R. C. Friedman, et al., eds., *Sex Differences in Behavior,* Wiley, New York, 1974, pp. 413–431.

43. See Lamb, M. E., "Physiological mechanisms in the control of maternal behavior in rats: A review," *Psychological Bulletin* 82(1) (1975): 104–119.

44. See Little, B. R., "Psychospecialization: Functions of differential interests in persons and things," *Bulletin of the British Psychological Society* 21 (1968): 113. Also Linesley, W. J., and Bromley, D. B., *Person Perception in Childhood and Adolescence,* Wiley, London, 1973, and Maccoby, E. M., and Jacklin, C. N., *Psychology of Sex Differences,* Stanford University Press, Stanford, 1974, and Goodenough, E. W., "Interest in persons as an aspect of sex difference in the early years," *Genetic Psychology Monographs* 55 (1957): 287–323.

45. See Lynn, D. B., *The Father: His Role in Child Development,* Brooks Cole, Monterey, California, 1974, pp. 14–21, for a review.

46. See Peck, E., and Senderowitz, J., eds., *Pronatalism: The Myth of Mom and Apple Pie,* Crowell, New York, 1974.

47. See Money, J., and Ehrhardt, A. A., *Man and Woman, Boy and Girl,* Johns Hopkins Press, Baltimore, 1972, for a comprehensive treatment.

48. Money, J., and Lewis, V. G., "IQ, genetics and accelerated growth: Adrenogenital syndrome," *Bulletin of the Johns Hopkins Hospital* 118 (1966): 365–373.

49. Ehrhardt, A. A., and Baker, S. W., "Fetal androgens, human central nervous system differentiation, and behavior sex differences," in R. C. Friedman et al., eds., *Sex Differences in Behavior*, pp. 33–51.

50. Baker, S. W., and Ehrhardt, A. A., "Prenatal androgen, intelligence, and cognitive sex differences," in R. C. Friedman et al., eds., *Sex Differences in Behavior,* pp. 53–76.

51. Descriptions of all these cases appear in Money, J., and Ehrhardt, A. A., *Man and Woman, Boy and Girl,* Johns Hopkins University Press, Baltimore, 1972.

52. See Stoller, R. J., *Sex and Gender: On the Development of Masculinity and Femininity,* Science House, New York, 1968.

53. Diamond, M., "A critical evaluation of the ontogeny of human sexual behavior," *Quarterly Review of Biology* 40 (1965): 147–175.

54. Morris, J., *Conundrum,* Signet, New York, 1975.

55. Ibid., p. 25–26.

56. See Benjamin, H., *The Transsexual Phenomenon,* Julian Press, New York, 1966. Also Person, E. S., and Onesey, L., "The psychodynamics of male transsexualism," in R. C. Friedman et al., eds., *Sex Differences in Behavior,* pp. 315–326. Also Green, R., *Sexual Identity Conflict in Children and Adults,* Penguin, Baltimore, 1975.

57. Barlow, D. H., Reynolds, E. H., and Agras, W. S., "Gender identity change in a transsexual," *Archives of General Psychiatry* 28(4) (1973): 569–579.

CHAPTER 2. The Psychological Maintenance System

1. Lewis, M., "Early sex differences in the human: Studies of socioemotional development," *Archives of Sexual Behavior* 4(4) (1975): 329–335.

2. Rheingold, H. L., and Cook, K. V., "The content of boys' and girls' rooms as an index of parents' behavior," *Child Development* 46(2) (1975): 459–463.

3. Ibid., p. 461.

4. Van Gelder, L., and Carmichael, C., "But what about our sons?", *Ms.* October 1975, pp. 52–56, 94–95.

5. Rothbart, M. K., and Maccoby, E. E., "Parents' differential reactions to sons and daughters," *Journal of Personality and Social Psychology* 4 (1966): 237–243.

6. Seavey, C. A., Katz, P. A., and Zalk, S. R., "Baby X: The effect of gender labels on adult responses to infants," *Sex Roles* 1(2) (1975): 103–109.

7. Gurwitz, S. B., and Dodge, K. A., "Adults' evaluations of a child as a function of sex of adult and sex of child," *Journal of Personality and Social Psychology* 32(5) (1975): 822–828.

8. See Korner, A. F., "Methodological considerations in studying sex differences in the behavioral functioning of newborns," in Friedman, R. C., Richart, R. M., and Vande Wiele, R. L., eds., *Sex Differences in Behavior*, Wiley, New York, 1974, pp. 197-208. See also other articles in this excellent volume cited below.

9. Korner, A. F., "Sex differences in newborns with special reference to differences in the organization of oral behavior," *Journal of Child Psychology and Psychiatry* 14 (1973): 19–29.

10. Moss, H. A., "Early sex differences and mother-infant interaction," in Friedman, et al., eds., *Sex Differences in Behavior*, pp. 149–164.

11. Lewis, M., and Weinraub, M., "Sex of parent x sex of child: Socioemotional development," in Friedman, et al., eds., *Sex Differences in Behavior*, pp. 165–190.

12. Ibid., p. 170.

13. Rosenblum, L. A., "Sex differences in mother-infant attachment in monkeys," in Friedman, et al., eds., *Sex Differences in Behavior*, pp. 123–142.

14. See Lewis, M., "State as an infant-environment interaction: An analysis of mother-infant interaction as a function of sex," *Merrill-Palmer Quarterly* 18(2) (1972): 95–121.

15. Maccoby, E. E., and Jacklin, C. G., *The Psychology of Sex Differences*, Stanford University Press, Stanford, 1974, p. 348. See also Sherman, J. A., *On the Psychology of Women: A Survey of Empirical Studies,* Thomas, Springfield, Illinois, 1971, for a comprehensive survey of psychological studies of women.

16. Sears, R. R., "Development of gender role," in F. A. Beach, ed., *Sex and Behavior,* Wiley, New York, 1965, p. 159 (quoted by Maccoby and Jacklin, P. 339).

17. "The Spocks: Bittersweet recognition in a revised classic," *New York Times,* March 19, 1976, p. 28 (quote is from preface of *Baby and Child Care,* 3rd ed., 1976).

18. Kagan, J., "The concept of identification," *Psychological Review* 65(5) 1958): 296–305.

19. Freud, S., *An Outline of Psychoanalysis,* Standard edition trans. J. Strachey, Hogarth Press, Ltd., London, 1940, p. 188, quoted in Mitchell, J., *Psychoanalysis and Feminism,* Pantheon, New York, 1974, p. 49–50.

20. Freud, S., "Some psychical consequences of the anatomical distinction between the sexes," Standard edition, trans. J. Strachey, Hogarth Press, London (originally published in 1925), in Strouse, J., ed., *Women and Analysis,* Grossman Publishers (Viking), 1974, p. 21.

21. Freud, S., "Femininity," *New Introductory Lectures on Psychoanalysis,* trans. J. Strachey, Norton, 1933, reprinted in Strouse, J., ed., *Women and Analysis*, p. 92.

22. Freud, S., "Three essays on sexuality," Standard edition, trans. J. Strachey, Hogarth Press, London, Vol. 7, 1905, pp. 219–220, quoted by Mitchell, J., "On Freud and the distinction between the sexes," in Strouse, J., ed., *Women and Analysis,* p. 31.

23. Freud, S., "Femininity," *New Introductory Lectures on Psychoanalysis,* trans. J. Strachey, Norton, 1933, reprinted in Strouse, J., *Women and Analysis,* p. 76.

24. Deutsch, H., *The Psychology of Women Volume 1 – Girlhood,* Bantam, New York, 1973, (originally published in 1944), p. 298.

25. Ibid., p. 394.

26. Lundberg, F., and Farnham, M. F., *Modern Woman: The Lost Sex,* Universal Library, New York, 1947.

27. Roiphe, A., "What women psychoanalysts say about women's liberation," *New York Times Magazine,* February 13, 1972, p. 63.

28. Horney, K., "The flight from womanhood," *International Journal of Psychoanalysis* 7 (1926): 324–339. Also appears in Miller, J. B., ed., *Psychoanalysis and Women,* Penguin, Baltimore, 1973, pp. 5–20.

29. Horney, in Miller, p. 19.

30. Thompson, C., "'Penis envy' in women," *Psychiatry* 6 (1943): 123–25. Also appears in Miller, *Psychoanalysis and Women,* pp. 52–57.

31. See Koedt, A., "The myth of the vaginal orgasm," in Koedt, A., Levine, E., and Rapine, A., eds., *Radical Feminism,* Quadrangle, New York, 1973, pp. 198–207.

32. Masters, W. H., and Johnson, V. E., *Human Sexual Response,* Little, Brown, Boston, 1966.

33. Degler, C. N., "What ought to be and what was: Women's sexuality in the nineteenth century," *American Historical Review,* (1974): 1467-1490.

34. Mitchell, J., *Psychoanalysis and Feminism,* Pantheon, New York, 1974.

35. See Mischel, W., "A social learning view of sex differences in behavior," in E. E. Maccoby, ed., *The Development of Sex Differences,* Stanford University Press, Stanford, 1966, pp. 56–81. See also Mischel, W., "Sex typing and socialization," in P. H. Mussen, ed., *Carmichael's Manual of Child Psychology,* 3rd ed., Wiley, New York, 1970, Volume II, pp. 3–72.

36. Bandura, A., "Influence of models' reinforcement contingencies on the acquisition of imitative responses," *Journal of Personality and Social Psychology* 1 (1965): 589–595.

37. Kagan, J., "Acquisition and significance of sex typing and sex role identity," in M. L. Hoffman and L. W. Hoffman, eds., *Review of Child Development Research*, Volume I, Russell Sage Foundation, New York, pp.137–167.

38. Ibid., p. 138.

39. Ibid., p. 147.

40. Ibid., p. 137.

41. Emmereich, W., "Socialization and sex role development," in *Life-Span Developmental Psychology,* Academic Press, New York, 1973, pp. 123–144.

42. Kohlberg, L., and Ullman, D. Z., "Stages in the development of psychosexual concepts and attitudes," in Friedman et al., *Sex Differences in Behavior,* pp. 209–222. An earlier version, Kohlberg, L., "A cognitive-developmental analysis of children's sex-role concepts and attitudes," is in Maccoby, E. E., *The Development of Sex Differences,* pp. 82–172.

43. Ibid., p. 210.

44. Money, J., Hampson, J., and Hampson, J., "Imprinting and the establishment of gender role," *Archives of Neurology and Psychiatry* 77 (1957): 333-336.

45. Lynn, D. B., *Parental and Sex-role Identification,* McCutchan Publishing Corporation, Berkeley, California, 1969.

46. See Harrison, B. G., *Unlearning the Lie: Sexism in School,* Morrow, New York, 1974. See also Stacey, J., Bereaud, S., and Daniels, J., eds., *And Jill Came Tumbling After: Sexism in American Education,* Dell, New York, 1974; Sexton, P., *The Feminized Male,*

Random House, New York, 1970; Austin, D. E., Clark, V. B., and Fitchett, G. W., *Reading Rights for Boys: Sex Role in Language Experience,* Appleton-Century-Crofts, New York, 1971; and *Dick and Jane as Victims: Sex Stereotyping in Children's Readers,* Women on Words and Images, Princeton, New Jersey, 1972.

47. Harrison, B. G., "Feminist experiment in education," in Stacey, et al., eds., *And Jill Came Tumbling After*, p. 380.

48. Cherry, L., "Teacher-child verbal interaction: An approach to the study of sex differences," in B. Thorne and N. Henley, eds., *Language and Sex: Difference and Dominance,* Newbury House, Rowley, Massachusetts, 1975, pp. 172–183.

49. Bowerman, C. E., and Kinch, J. W., "Changes in family and peer orientation of children between the 4th and 10th grade," *Social Forces,* 37(3) (1959): 206–211.

50. Coleman, J. S., *The Adolescent Society,* Free Press, Glencoe, Illinois, 1961.

51. Douvan, E., "Independence and identity in adolescence," *Children* 4 (1957): 186–190. See also Douvan, E., "Sex differences in adolescent character processes," *Merrill-Palmer Quarterly* 6 (1960): 203–211; Douvan, E., and Adelson, J., *The Adolescent Experience,* Wiley, New York, 1966, and Konopka, G., *Young Girls: A Portrait of Adolescence,* Prentice-Hall, Englewood Cliffs, 1976.

52. See Komarovsky, M., *Blue-Collar Marriage,* Vintage, New York, 1967, for a study of sexually divided peer groups in the working class.

53. Fisher, E., "Children's books: The second sex, junior division," in Stacey, et al., eds., *And Jill Came Tumbling After*, pp. 116–122. See also references in note 46.

54. "A report on children's toys," from *Ms.* magazine, in Stacey, et al., eds., *And Jill Came Tumbling After*, pp. 123–125.

55. Sternglanz, S. H., and Serbin, L. A., "Sex role stereotyping in children's television programs," *Developmental Psychology* 10(5) (1974): 710–715.

56. See Komisar, L., "The image of woman in advertising," in Gornick, V., and Moran, B. K., eds., *Woman in Sexist Society,* New American Library, 1972, pp. 304–317. (Several other articles in this collection are also relevant.) See also Franzwa, H. H., "Female roles in women's magazine fiction, 1940–1970," in R. K. Unger and F. L. Denmark, eds., *Woman: Dependent or Independent Variable?,* Psychological Dimensions, New York, 1975.

57. See Chesler, P., *Women and Madness*, Avon Books, New York, 1972.

58. Broverman, I. K., Broverman, D. M., Clarkson, F. E., Rosenkrantz, P. S., and Vogel, S. R., "Sex role stereotypes and clinical judgments of mental health," *Journal of Consulting and Clinical Psychology* 34(1) (1970): 1–7. See also Franks, V., and Burtle, V., eds., *Women in Therapy*, Brunner-Mazel, New York, 1974.

59. Constantinople, A., "Masculinity-femininity an exception to a famous dictum," *Psychological Bulletin* 80(5) (1973): 389–407.

60. Terman, L., and Miles, C. C., *Sex and Personality,* McGraw-Hill, New York, 1936.

61. Maccoby, E. E., "Sex differences in intellectual functioning," in *The Development of Sex Differences,* Stanford University Press, Stanford, California, 1966, pp. 25–55.

62. Broverman, D. M., Klaiber, E. L., Kobayashi, Y., and Vogel, W., "Roles of activation and inhibition in sex differences in cognitive abilities," *Psychological Review* 75(1) (1968): 23–50.

63. Singer, G., and Montgomery, R. B., "Comment on roles of activation and inhibition in sex differences in cognitive abilities," *Psychological Review* 76(3) (1969): 325–327. See also "Reply" to above by Broverman, et al., *Psychological Review* 76(3) (1969): 328–331.

64. Parlee, M. B., "Comments on 'Roles of activation and inhibition in sex differences in cognitive abilities' by Broverman, et al.," *Psychological Review* 79(2) (1972): 180–184.

65. Ibid., p. 180.

66. Gray, J. A., and Buffery, A. W. H., "Sex differences in emotional and cognitive behaviour in mammals including man: Adaptive and neural bases," *Acta Psychologica* 35 (1971): 89–111. See also Gray, J. A., "Sex differences in emotional behaviour in mammals including man: Endocrine bases," *Acta Psychologica* 35 (1971): 29–46.

67. Archer, J., "Sex differences in emotional behaviour: A reply to Gray and Buffery," *Acta Psychologica* 35 (1971): 415–429.

68. Witkin, H. A., Dyk, R. B., Faterson, H. F., Goodenough, D. R., and Karp, S. A., *Psychological Differentiation,* Wiley, New York, 1962. See also Kagan, J., and Kogan, N., "Individuality and cognitive performance," in Mussen, P. H., ed., *Carmichael's Manual of Child Psychology,* Volume I, Wiley, New York, 1970, pp. 1273–1365, especially pp. 1323–1342.

69. Coates, S., "Sex differences in field independence among preschool children," in Friedman, et al., eds., *Sex Differences in Behavior*, pp. 259–274.

70. McClelland, D. C., "Wanted: A new self-image for women,"

in R. J. Lifton, ed., *The Woman in America,* Beacon Press, Boston, 1967, p. 181.
71. Silverman, J., "Attentional styles and the study of sex differences," in D. I. Mostofsky, ed., *Attention: Contemporary Theory and Analysis,* Appleton-Century-Crofts, New York, 1970, pp. 61–98. The references in this paragraph (notes 71-73) were drawn from Kogan, N., "Sex differences in creativity and cognitive styles," paper presented at the Invitational Conference on Cognitive Styles and Creativity in Higher Education, Montreal, 1972. Grateful acknowledgment is hereby given to Nathan Kogan for use of the paper.
72. Helson, R., "Sex differences in creative style," *Journal of Personality* 35 (1967): 214–233. See also two later articles by Helson: *Journal of Personality* 36 (1968): 33–48 and *Journal of Personality* 38 (1970): 344–363.

73. Gutmann, D., "Female ego styles and generational conflict," in Bardwick, J. M., et al., eds., *Feminine Personality and Conflict,* Brooks Cole, Belmont, California, 1970, pp. 77–96.

74. Maccoby, E. E., and Jacklin, C. N., *The Psychology of Sex Differences,* Stanford University Press, Stanford, 1974, pp. 63–133.

75. See Bock, D. R., and Kolakowski, D., "Further evidence of sex-linked major-gene influence in human spatial visualizing ability," *American Journal of Human Genetics* 25 (1973): 1–14. See also discussion in Maccoby and Jacklin, pp. 120–122.

76. Helson, R., "Sex differences in creative style," *Journal of Personality* 35 (1967): 214–233.

77. Osen, L. M., *Women in Mathematics,* The MIT Press, Cambridge, 1974.

78. Ibid., p. 165.

79. Ibid., p. 104 (quoted from Tabor, M. E., *Pioneer Women,* The Sheldon Press, London, 1933, p. 107).

80. Maccoby and Jacklin, p. 75.

81. Ibid., p. 125–127.

82. Levy-Agresti, J., "Ipsilateral projection systems and minor hemisphere function in man after neuromissurotomy," *Anatomical Record* 61 (1968): 1151.

83. Maccoby and Jacklin, p. 126.

84. Woolf, V., *A Room of One's Own*, Harcourt, Brace & World, New York, 1957 (originally published in 1929), p. 49–50.

85. Kogan, N., "Creativity and sex differences," *The Journal of Creative Behavior* 8(1) (1975): 1–14.

86. Maccoby and Jacklin, pp. 169–226.

87. Parsons, T., and Bales, R. F., eds., *Family, Socialization and Interaction Process,* Free Press, New York, 1955. See also Slater, P., "Parental role differentiation," *American Journal of Sociology* 67(3) (1961): 296–311.

88. Aries, E., "Interaction patterns and themes of male, female and mixed groups," a paper presented at the American Psychological Association Convention, New Orleans, September, 1974.

89. Barry, H. III, Bacon, M. K., and Child, I. I., "A cross-cultural survey of some sex differences in socialization," *Journal of Abnormal and Social Psychology* 55 (1957): 327–332.

90. Klein, V., *The Feminine Character: History of An Ideology,* University of Illinois Press, 1975, p. 164.

91. Fasteau, M. F., *The Male Machine,* McGraw-Hill, New York, 1974, p. 1.

CHAPTER 3. The Social Maintenance System: The Family

1. Mair, L., *Marriage*, Penguin, Baltimore, 1971, p. 8.

2. Linton, R., *The Study of Man,* Appleton-Century, New York, 1936, p. 175.

3. Cited in Gardner, H., *The Quest for Mind: Piaget, Lévi-Strauss and the Structuralist Movement,* Knopf, New York, 1973, p. 125. (Gardner reports the vignette originally appeared in Mead, M., *Sex and Temperament in Three Primitve Societies,* Morrow, New York, 1935.)

4. Hart, C. W. M., and Pilling, A. R., *The Tiwi of North Australia,* Holt, Rinehart & Winston, New York, 1960, p. 16.

5. Goodale, J. C., *Tiwi Wives,* University of Washington Press, Seattle, 1971. See also Rohrlich-Leavitt, R., Sykes, B., and Weatherford, E., "Aboriginal woman: Male and female anthropological perspectives," in Reiter, R., ed., *Toward An Anthropology of Women,* Monthly Review Press, New York, 1975, pp. 110–126.

6. Rubin, G., "The traffic in women," in Reiter, p. 177.

7. Brownmiller, S., *Against Our Will: Men, Women and Rape,* Simon and Schuster, New York, 1975, p. 16.

8. Ibid., p. 17.

9. Gough, K., "The origin of the family," *Journal of Marriage and the Family* 33(4) (1971). This article has also been reprinted in

numerous anthologies on the family and sex roles (e.g., Perrucci, C. C., and Targ, D. B., eds., *Marriage and the Family,* McKay, New York, 1974, pp. 71–93).

10. Gough, in Perrucci and Targ, p. 86.

11. Ibid., p. 73.

12. Friedl, E., *Women and Men: An Anthropologist's View,* Holt, Rinehart & Winston, New York, 1975, pp. 31–32.

13. Draper, P., "!Kung women: Contrasts in sexual egalitarianism in foraging and sedentary contexts," in Reiter, pp. 77–109.

14. Rogers, S. C., "Female forms of power and the myth of male dominance: A Model of female/male interaction in peasant society," *American Ethnologist* 2(4) (1975): 727–756. See also Cornelisen, A., *Women of the Shadows,* Little, Brown & Co., Boston 1976, for an excellent account of female life in Southern Italy.

15. Reiter, p. 282.

16. Friedl, E., *Women and Men: An Anthropologist's View,* Holt, Rinehart & Winston, New York, 1975.

17. Shorter, E., *The Making of the Modern Family,* Basic Books, New York, 1975.

18. Ibid., p. 55.

19. Ibid., p. 259.

20. Capellanus, A., *The Art of Courtly Love,* trans. J. J. Parry, Frederick Ungar Publishing Co., New York, 1970, p. 100.

21. Shorter, p. 168.

22. Ibid., pp. 203–204.

23. Ibid., p. 205.

24. Ibid.

25. Engels, F., *The Origin of the Family, Private Property and the State,* International Publishers, New York, 1942, p. 65.

26. Tripp, M., "The free married woman," in *Woman in the Year 2000,* M. Tripp, ed., Arbor House, New York, p. 53.

27. The following sources contain much information on the changing status of the working woman in America: Sweet, J. A., *Women in the Labor Force*, Seminar Press (Harcourt Brace Jovanovich), New York, 1973. Kreps, J. M., ed., *Women and the American Economy: A Look to the 1980s,* Prentice-Hall, Englewood Cliffs, 1976. Smuts, R. W., *Women and Work in America*, Schocken, New York, 1971. Kreps, J., *Sex in the Marketplace: American Women at*

Work, Johns Hopkins Press, Baltimore, 1971. Kreps, J., and Clark, R., *Sex, Age, and Work: The Changing Composition of the Labor Force,* Johns Hopkins Press, Baltimore, 1975. Brownlee, W. E., and Brownlee, M. M., eds., *Women in the American Economy: A Documentary History*, Yale University Press, New Haven, 1976.

28. U. S. Department of Labor, Manpower Administration report, 1967, cited in Sweet, p. 30.

29. See Hoffman, L. W., and Nye, F. I., *Working Mothers,* Jossey-Bass, San Francisco, 1974. A good nonacademic treatment is Curtis, J., *Working Mothers,* Doubleday, New York, 1976.

30. Ibid., p. 164.

31. Scanzoni, J. H., *Sex Roles, Lifestyles and Childbearing: Changing Patterns in Marriage and the Family,* Free Press, New York, 1975, p. 134.

32. Waite, L. J., and Stolzerberg, R. M., "Intended childbearing and labor force participation of young women: Insights from nonrecursive models," *American Sociological Review* 41(2) (1976): 235–251.

33. Gilman, C. P., *Women and Economics,* Source Book Press (originally Small, Maynard and Co., Boston, 1898), pp. 329–330.

34. Vanek, J., "Time spent in housework," *Scientific American* 231(5) (1974): 116–120.

35. Reported in *New York Times,* April 27, 1976, based on Census Bureau study.

36. McClelland, D., *The Achieving Society,* Free Press, New York, 1967.

37. See Deaux, K., and Emswiller, T., "Explanations of successful performances on sex-linked tasks: What is skill for the male is luck for the female," *Journal of Personality and Social Psychology* 29 (1974): 846–855. See also, for a general discussion: Deaux, K., *The Behavior of Women and Men,* Brooks Cole, Monterey, California, 1976, especially Chapters 3 and 4.

38. Horner, M. S., "Femininity and successful achievement: A basic inconsistency," in Bardwick, J. M., et al., *Feminine Personality and Conflict,* Brooks Cole, Belmont, California, 1970, pp. 45–76.

39. Ibid., p. 61.

40. Konopka, G., *Young Girls: A Portrait of Adolescence,* Prentice-Hall, Englewood Cliffs, 1976, p. 15–16.

41. Bailyn, L., "Notes on the role of choice in the psychology of

professional women," in R. J. Lifton, ed., *The Woman in America,* Beacon Press, Boston, 1964, p. 238.

42. Lopate, C., *Women in Medicine,* Johns Hopkins Press, Baltimore, 1968.

43. Herzfeld, M., A study of changes in self-concept at Radcliffe, 1966, unpublished (personal communication, 1969).

44. Ginzberg, E., *Life Styles of Educated Women,* Columbia University Press, New York, 1966.

45. Williams, P., Study of Radcliffe graduates who went to medical school, Radcliffe Institute, unpublished (personal communication, 1969).

46. Rossi, A. J., "Barriers to the career choice of engineering, medicine or science among American women," in J. A. Mattfield and C. G. Van Aken, eds., *Women and the Scientific Professions,* MIT Press, Cambridge, 1965.

47. Ibid., p. 83.

48. Ginzberg, p. 30.

49. Bernard, J., *Academic Women,* Penn State University Press, University Park, Pennsylvania, 1964.

50. Rossi, p. 98.

51. Ibid., p. 89.

52 Ginzberg, p. 20.

53. Rossi, p. 105.

54. Dodge, N. T., *Women in the Soviet Economy,* Johns Hopkins Press, Baltimore, 1966.

55. Rossi, p. 54.

56. Fasteau, M. F., *The Male Machine,* McGraw-Hill, New York, 1974. See also Komarovsky, M., *Dilemmas of Masculinity,* Norton, New York, 1976.

57. See Kanowitz, L., *Women and the Law,* University of New Mexico Press, Albuquerque, 1969, for a good discussion of the law and women.

58. See Lynn, D. B., *The Father: His Role in Child Development,* Brooks Cole, Monterey, California, 1974. Good nonacademic treatments are: Biller, H., and Meredith, D., *Father Power,* Anchor Books, New York, 1975, and Green, M., *Fathering,* McGraw-Hill, New York, 1976.

CHAPTER 4. The Social Maintenance System: Symbolism

1. Steiner, F., *Taboo*, Penguin, Baltimore, 1967 p. 147.

2. Firth, R., *Symbols, Public and Private,* Cornell University Press, Ithaca, New York, 1973.

3. See Weideger, P., *Menstruation and Menopause: The Physiology and Psychology, the Myth and the Reality.* Knopf, New York, 1976, for a good discussion.

4. Douglas, M., *Purity and Danger: An Analysis of Concepts of Pollution and Taboo,* Penguin, Baltimore, 1970, p. 174.

5. Remy, N., *Demonolatry,* 1595, English translation by E. A. Ashwin, London, 1930, p. 56 (quoted in Parrinder, G., *Witchcraft: European and American,* Faber and Faber, London, 1963, p. 109.)

6. Goody, E., "Legitimate and illegitimate aggression in a West African state," pp. 242–3, in M. Douglas, ed., *Witchcraft Confessions and Accusations,* Tavistock, New York, 1970, pp. 207–244.

7. Douglas, M., *Purity and Danger,* p. 120.

8. Spooner, B., "The evil eye in the Middle East," p. 315, in Douglas, M., *Witchcraft Confessions and Accusations,* pp. 311–319.

9. Parrinder, G., p. 192.

10. Ibid., p. 113.

11. Douglas, M., *Purity and Danger,* p. 120.

12. Faithorn, E., "The concept of pollution among the Káfe of the Papua New Guinea Highlands," pp. 127–140, in Reiter, R. R., ed., *Toward an Anthropology of Women*, Monthly Review Press, New York, 1975.

13. Douglas, M., *Purity and Danger,* p. 124.

14. Leach, E., *Claude Lévi-Strauss,* Viking, New York, 1970, p. 54.

15. See Fontenrose, J., *The Ritual Theory of Myth,* University of California Press, Berkeley, 1966, for a discussion.

16. Eliade, M., *Myths, Dreams and Mysteries,* Harper & Row, New York, 1967, p. 172.

17. Hall, C. S., and Lindzey, G. *Theories of Personality,* 2nd edition, Wiley, New York, 1970, p. 83.

18. Jung, C. G., et al., *Man and His Symbols,* Doubleday, New York, 1964, p. 177.

19. *I Ching,* trans. John Blofeld, Dutton, New York, 1968, pp. 90–91.

20. Cooper, J. C., *Taoism: The Way of the Mystic*, Weiser, New York, 1972, p. 28.

21. *I Ching*, pp. 85–86.

22. Eliade, p. 174–175.

23. Rawson, P., *Tantra: The Indian Cult of Ecstasy*, Avon, New York, 1973, p. 14.

24. Lederer, W., *The Fear of Women*, Harcourt Brace Jovanovich, New York, 1968, p. 136.

25. Eliade, M., p. 184.

26. Neumann, E., *The Great Mother: An Analysis of an Archetype*, trans. R. Manheim, Princeton University Press, Princeton, 1972, p. 149.

27. See Finley, M., "Archaeology and History," *Daedalus* 100 (1971): 168–86 (see Pomeroy).

28. Ucko, P. J., *Anthropomorphic Figurines of Predynastic Egypt and Neolithic Crete*, Royal Anthropological Institute, London, 1968, p. 316, quoted in Pomeroy, p. 14.

29. Pomeroy, S. B., *Goddesses, Whores, Wives, and Slaves: Women in Classical Antiquity*, Schocken, New York, 1975, p. 14.

30. Eliade, p. 166.

31. Scholem, G., *On the Kabbalah and Its Symbolism*, trans. R. Manheim, Schocken, New York, 1969, p. 108.

32. Ibid., p. 107.

33. See Power, E., *Medieval Women*, ed., M. M. Postan, Cambridge University Press, Cambridge, 1975, p. 19.

34. Pomeroy, p. 8.

35. Luther, M., "The natural place of women," in Verene, D. P., ed., *Sexual Love and Western Morality*, Harper & Row, New York, 1972, pp. 134–135.

36. Kirk, G. S., *Myth: Its Meaning and Functions*, Cambridge University Press, Cambridge, 1971, p. 230.

37. Campbell, J., *The Hero With a Thousand Faces*, Meridian, New York, 1956.

38. Rank, O., *The Myth of the Birth of the Hero*, trans. F. Robbins and S. E. Jellife, Brunner, New York, 1952, p. 61.

39. See Vanggaard, T., *Phallos: A Symbol and Its History in the Male World*, International Universities Press, New York, 1974.

40. Eibl-Eibesfelt, E., "Similarities and differences between cultures in expressive movements," in *Nonverbal Communication*, R. A. Hinde, ed., Cambridge University Press, Cambridge, 1972, pp. 297–312.

41. Van Gennep, A., *The Rites of Passage*, trans. M. B. Vizedom and G. L. Caffee, University of Chicago Press, 1960.

42. Bettelheim, B., *Symbolic Wounds: Puberty Rites and the Envious Male*, Collier Books, New York, 1962.

43. Whiting, J. W. M., Kluckhohn, R., and Anthony, A., "The functions of male initiation ceremonies at puberty," in *Readings in Social Psychology*, E. E. Maccoby, T. M. Newcomb, E. L. Hartley, eds., Holt, Rinehart & Winston, New York, 1958, pp. 359–370.

44. Ibid., p. 370.

45. Parker, S., Smith, J., and Ginat, J., "Father absence and cross-sex identity: The puberty rites controversy revisited," *American Ethnologist* 2(4) (1975): 687–706.

46. Eliade, M., *Rites and Symbols of Initiation: The Mysteries of Birth and Rebirth*, trans. W. R. Trask, Harper & Row, New York, 1958, pp. 131–132.

47. Bettelheim, *Symbolic Wounds*, pp. 128–130.

48. Hayes, R. O., "Female genital mutilation, fertility control, women's roles, and the patrilineage in modern Sudan: A functional analysis," *American Ethnologist* 2(4) (1975): 617–633.

49. Ibid., p. 624.

50. Brownmiller, S., *Against Our Will: Men, Women and Rape*, Simon and Schuster, New York, 1975.

51. Lederer, pp. 180–181.

52. See Mair, L., *Marriage*, Penguin, 1971, especially pp. 48–73.

53. Ibid., p. 58.

54. Mason, J. P., "Sex and symbol in the treatment of women: The wedding rite in a Libyan oasis community," *American Ethnologist* 2(4) (1975): 649–661.

55. See Bodine, A., "Sex differentiation in language," in B. Thorne and N. Henley, eds., *Language and Sex: Difference and Dominance*, Newbury House, Rowley, Massachusetts, 1975, pp. 130–151, for a good discussion.

56. Ibid., p. 130.

57. Key, M. R., *Male/Female Language*, Scarecrow Press, Metuchen, New Jersey, 1975, p. 91.

58. Lakoff, R., *Language and Woman's Place,* Harper & Row, New York, 1975.

59. Ibid., p. 19.

60. Aries, E., "Interaction patterns and themes of male, female and mixed groups," a paper presented at the American Psychological Association convention, New Orleans, September, 1974.

61. Ibid.

62. Weitz, S., "Sex differences in nonverbal communication," *Sex Roles,* 2(2) (1976): 175–184.

63. Fishman, P., "Study of male-female conversations," paper presented at the American Sociological Association meeting, San Francisco, August, 1975.

64. Zimmerman, D. H., and West, C., "Sex roles, interruptions and silences in conversation," in B. Thorne and N. Henley, eds., *Language and Sex: Difference and Dominance*, Newbury House, Rowley, Massachusetts, 1975, pp. 105–129.

65. Ibid., p. 125.

66. Sachs, J., Lieberman, P., and Erickson, D., "Anatomical and cultural determinants of male and female speech," in R. Shuy and R. W. Fasold, eds., *Language Attitudes: Current Trends and Prospects,* Georgetown University Press, Washington, 1973, pp. 74–84. Also see Sachs, J., "Cues to the identification of sex in children's speech," in B. Thorne and N. Henley, eds., *Language and Sex: Difference and Dominance,* pp. 152–171.

67. Spacks, P. M., *The Female Imagination,* Knopf, New York, 1975, p. 77.

68. Friedan, B., *The Feminine Mystique*, Dell, 1963, p. 38.

69. Franzwa, H. H., "Female roles in women's magazine fiction," in Unger, R. K., and Denmark, F. L., eds., *Woman: Dependent or Independent Variable*, Psychological Dimensions, New York, 1975, pp. 42–53.

70. Nochlin, L., "Eroticism and female imagery in nineteenth century art," in T. B. Hess and L. Nochlin, eds., *Woman as Sex Object: Studies in Erotic Art,* 1730–1970, Art News Annual XXXVIII, Newsweek, New York, 1972, p. 10.

71. Berger, J., "The past seen from a possible future," *Selected Essays and Articles,* Penguin, 1972, p. 215 (quoted in Nochlin, p. 14).

72. Haskell, M., *From Reverence to Rape: The Treatment of Women in the Movies,* Penguin, Baltimore, 1973, p. 254.

73. Ibid., p. 255.

74. Mellon, J., *Women and Their Sexuality in the New Film,* Dell, New York, 1973, p. 23.

75. Ibid., p. 24.

76. Newcomb, H., *TV: The Most Popular Art,* Anchor Books, New York, 1974, pp. 48–49.

CHAPTER 5. Sex Role Change Through Space and Time

1. Smith, H., *The Russians,* Quadrangle, New York, 1976, p. 128.

2. Mandel, W. M., *Soviet Women*, Anchor, New York, 1975, p. 67.

3. Scott, H., *Does Socialism Liberate Women?* Beacon Press, Boston, 1974, p. 10.

4. Smith, p. 133.

5. Scott, p. 164.

6. Smith, p. 140.

7. Wong, A. K., "Women in China: Past and present," in Matthiasson, C. J., ed., *Many Sisters: Women in Cross-Cultural Perspective,* Free Press, New York, 1974, p. 230.

8. Curtin, K., *Women in China,* Pathfinder, New York, p. 13.

9. Ibid., p. 36.

10. Ibid., p. 51.

11. Ibid., p. 66.

12. Zborowski, M., and Herzog, E., *Life is With People,* Schocken, New York, 1962, p. 131.

13. Spiro, M. E., *Kibbutz – Venture in Utopia,* Harvard University Press, Cambridge, 1956, quoted in Rabin, A. I., "The sexes in the Israeli Kibbutz," in Seward, G. H., and Williamson, R. C., eds., *Sex Roles in Changing Society*, Random House, New York, 1970, p. 298.

14. Shafer, J., *The Reflection of Children's Sleeping Arrangements in the Social Structure of the Kibbutz,* Ichud, Tel Aviv, 1967. Quoted in Rabin; in Seward and Williamson, p. 298.

15. Rabin, in Seward and Williamson, p. 299.

16. *New York Times,* March 4, 1976, p. 26.

17. Ibid.

18. Stiller, N., "Peace without honor: The battle of the sexes in Israel," *Midstream*, May, 1976, p. 35.

19. Ibid., p. 36.

20. Ibid., p. 37.

21. Ibid., p. 38.

22. Ibid., p. 40.

23. Myrdal, A., and Klein, V., *Women's Two Roles*, Routledge & Kegan Paul, London, 1956.

24. Tiger, L., and Shepher, J., *Women in the Kibbutz*, Harcourt Brace Jovanovich, New York, 1975.

25. Anthony, K., *Feminism in Germany and Scandinavia*, Holt, New York, 1915, p. 221.

26. Moberg (1962) quoted in Liljestrom, R., "The Swedish Model," in Seward and Williamson, p. 203.

27. *Official Report to the United Nations on the Status of Women in Sweden*, Swedish Institute, Stockholm, 1968, p. 3.

28. Leijon, A. G., *Swedish Women–Swedish Men*, Swedish Institute, Stockholm, 1968, p. 145.

29. Ibid., p. 73.

30. Rossi, A. S., ed., *The Feminist Papers*, Columbia University Press, New York, 1973, p. 3.

31. Rossi, p. 42 (from Wollstonecraft) See Wollstonecraft, M., *A Vindication of the Rights of Women*, Norton, New York, 1967.

32. Ibid., pp. 57–8, (from Wollstonecraft).

33. Schneir, M., ed., *Feminism: The Essential Historical Writings*, Vintage, New York, 1972, p. 11 (from Wollstonecraft).

34. Rossi, p. 72, (from Wollstonecraft).

35. Schneir, p. 167 (from Mill). See Mill, J. S., *On the Subjection of Women*, Fawcett, New York, 1971.

36. Rossi, p. 203 (from Mill).

37. Ibid., p. 249.

38. Papachristou, J., *Women Together: A History in Documents of the Women's Movement in the United States*, Knopf, New York, 1976, p. 3.

39. Ibid., p. 16.

40. Ibid., p. 15.

41. Tocqueville, A. de, *Democracy in American,* Mentor, 1956, p. 234.

42. Sherrs, L., and Kazickas, J., *The American Women's Gazeteer,* Bantam Books, New York, 1976, p. 175.

43. Papachristou, p. 24–5.

44. Ibid., p. 27.

45. Ibid., p. 70.

46. Ibid.

47. Rossi, p. 355.

48. Papachristou, p. 76.

49. Ibid.

50. Ibid., p. 93.

51. Ibid., p. 96.

52. Ibid., p. 97.

53. Schneir, p. 237, p. 242 (from Gilman). See Gilman, C. P., *Women and Economics,* Source Book Press, New York, 1970 (originally published in 1898). Also *The Home,* University of Illinois Press, 1972 (originally published in 1903).

54. Chafe, W. H., *The American Woman,* Oxford University Press, New York, 1972, p. 92.

55. Ibid., p. 60.

56. Rossi, p. 533 (from Sanger).

57. Lundberg, F., and Farnham, M. F., *Modern Woman: The Lost Sex,* Universal Library, New York, 1947.

58. Friedan, B., *The Feminine Mystique,* Dell, New York, 1963.

59. de Beauvoir, S., *The Second Sex,* trans. H. M. Parshley, Knopf, New York, 1953.

60. Friedan, p. 364.

61. Guettel, C., *Marxism and Feminism,* Women's Press, Toronto, 1974, p. 3.

62. Yates, G. G., *What Women Want: The Ideas of the Movement,* Harvard University Press, Cambridge, 1975, p. 172. See also Freeman, J., *The Politics of Women's Liberation,* David McKay Company, Inc., New York, 1975.

63. Guettel, p. 39 (from Firestone). See Firestone, S., *The Dialectic of Sex,* Bantam, New York, 1971.

NAME INDEX

SUBJECT INDEX